Body-Art to Accompany

Human Anatomy and Physiology

SECOND EDITION

by
Solomon
Schmidt
Adragna

Saunders College Publishing
Philadelphia Fort Worth Chicago San Francisco
Montreal Toronto London Sydney Tokyo

BODY-ART to Accompany *Human Anatomy & Physiology, 2/E* by Solomon·Schmidt·Adragna

ISBN 0-03-046083-2

012 014 987654321

Body-Art to Accompany
HUMAN ANATOMY AND PHYSIOLOGY, second edition
by Solomon/Schmidt/Adragna

Preface

BODY-ART™ is an exclusive addition to the ancillary package accompanying *Human Anatomy and Physiology, 2/e* by Solomon/Schmidt/Adragna. Both instructors and students will find BODY-ART™ useful as an instructional or study aid. BODY-ART™ is available free to adopters of *Human Anatomy and Physiology, 2/e.*

What is BODY-ART™?

BODY-ART™ is a collection of 124 illustrations from the *Human Anatomy and Physiology, 2/e* textbook reproduced in black and white without their accompanying captions and labels. The pages of this book are perforated and three-hole punched for easy removal and storage. Each illustration is printed on only one side of a page to facilitate individual choice. These 124 figures are the same ones reproduced in color (with captions and labels) that are available as either Overhead Transparencies or 35mm Slides.

How to use BODY-ART™:

Because the figures appear without labels and captions, instructors can use BODY-ART™ in several ways. For example:

Lectures
Some instructors like to project transparencies of unlabeled or partially labeled figures during lecture sessions to assist them in motivating student interest. The BODY-ART™ figures can be used as transparency masters to create overhead transparencies that the instructor can then label and color to suit individual needs. Use of the BODY-ART™ in this way will give an instructor using *Human Anatomy and Physiology, 2/e* the widest range of teaching options now available with any textbook.

Notes
Each BODY-ART™ sheet also contains a small area at the bottom of the page for note-taking or caption writing, either by instructor or student. The student can follow the lecture and use of the corresponding transparency by the instructor, thus actively participating in the learning process.

Quizzes and Homework
Selected pages can be used for homework assignments, quizzes, or as components in examinations. Students' knowledge of organisms or processes can be evaluated by having them identify the pertinent unlabeled parts of those figures appearing in the BODY-ART™ collection. Or students could be asked to write their own captions for the illustrations.

Coloring Book
Each illustration consists of several parts to be colored. At the back of the book, there is a complete list of labels for the anatomical structures in the illustration which are to be identified, colored and learned. When a structure is colored on the illustration, the same color should be used to color the circle next to the corresponding label of that structure. Although there is not a set coloring scheme, convention is as follows:

> muscles - reddish brown
> ligaments - green
> arteries - bright red
> nerves - yellow

Once colored, the illustration becomes a reference for reviewing the structure prior to lab, lecture, or exam.

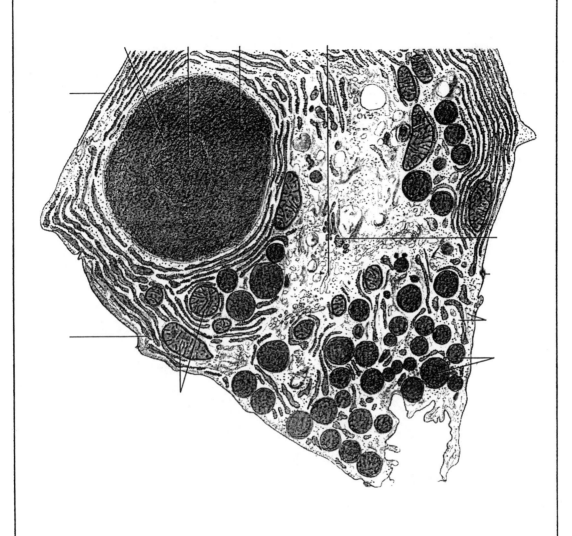

Figure 3-2

NOTES:

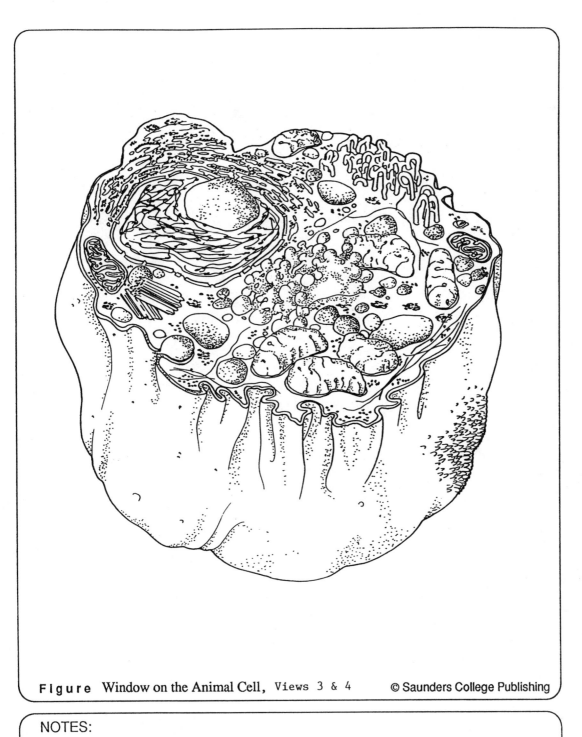

Figure Window on the Animal Cell, Views 3 & 4 © Saunders College Publishing

NOTES:

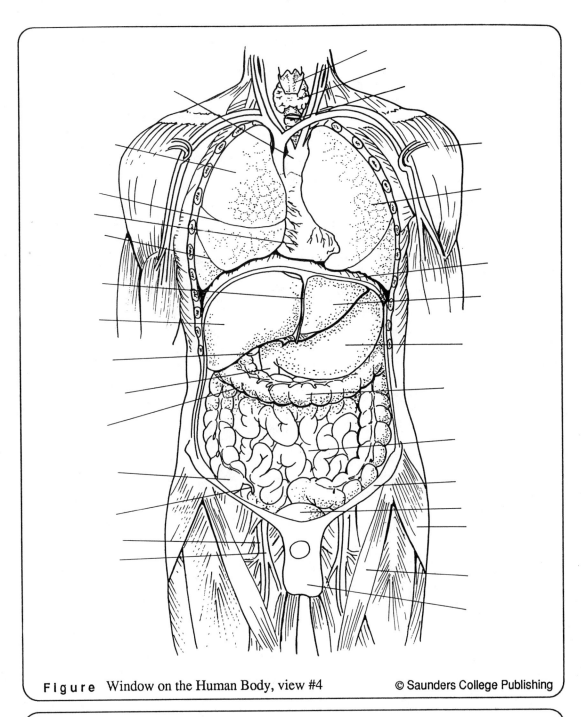

F i g u r e Window on the Human Body, view #4   © Saunders College Publishing

NOTES:

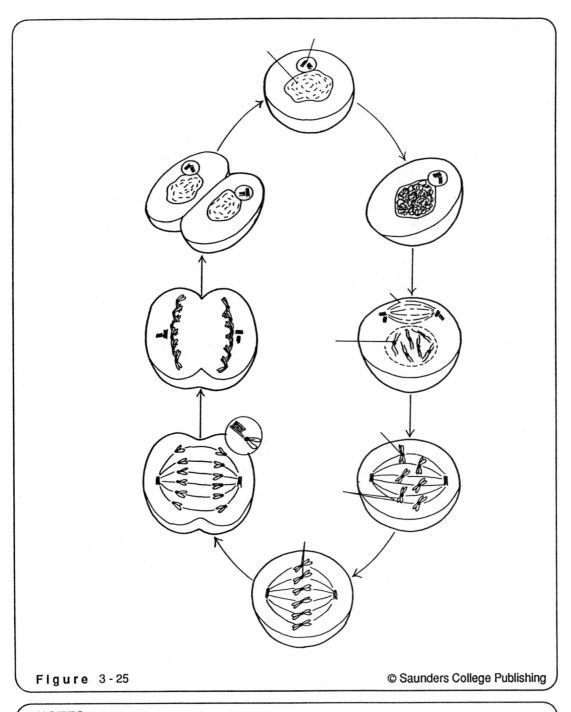

Figure 3-25

NOTES:

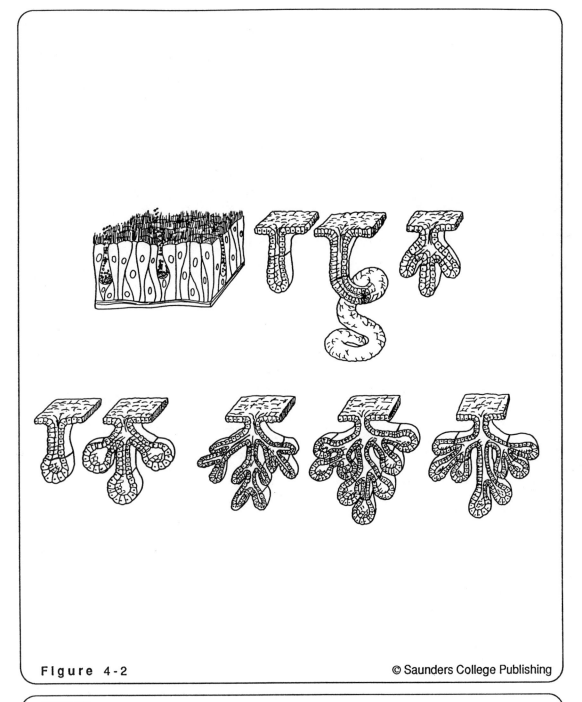

Figure 4-2

© Saunders College Publishing

NOTES:

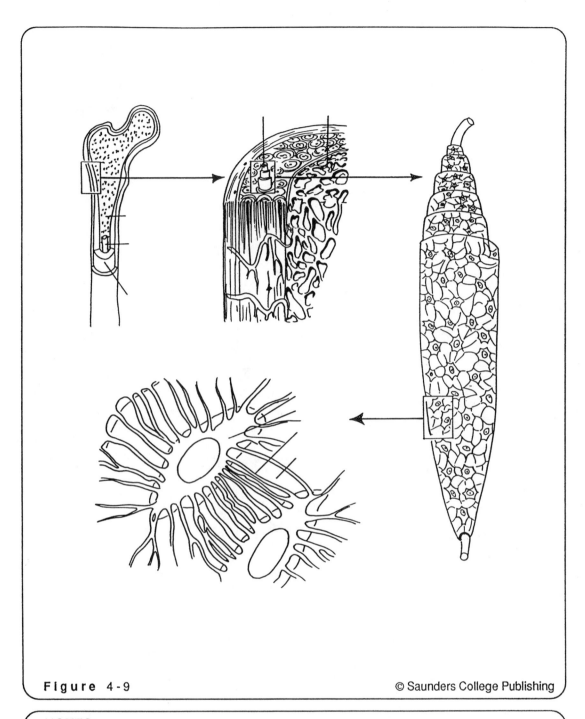

Figure 4-9

NOTES:

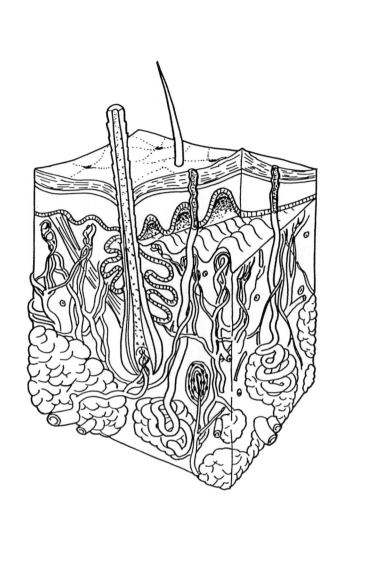

Figure 5-2

NOTES:

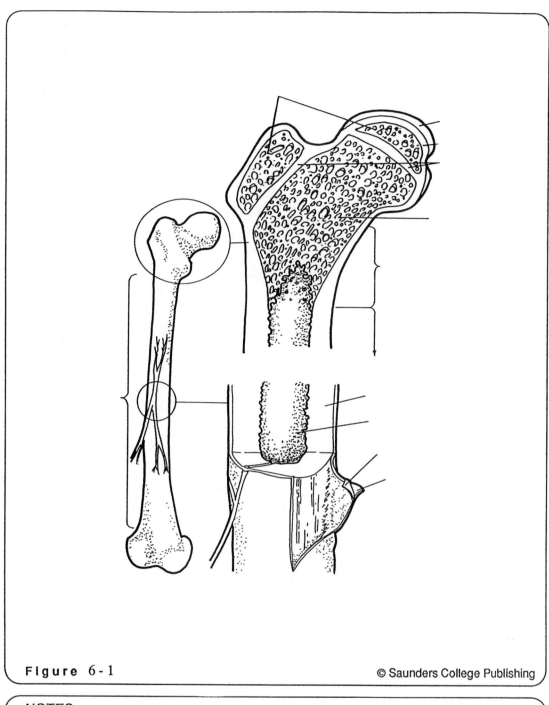

Figure 6-1

NOTES:

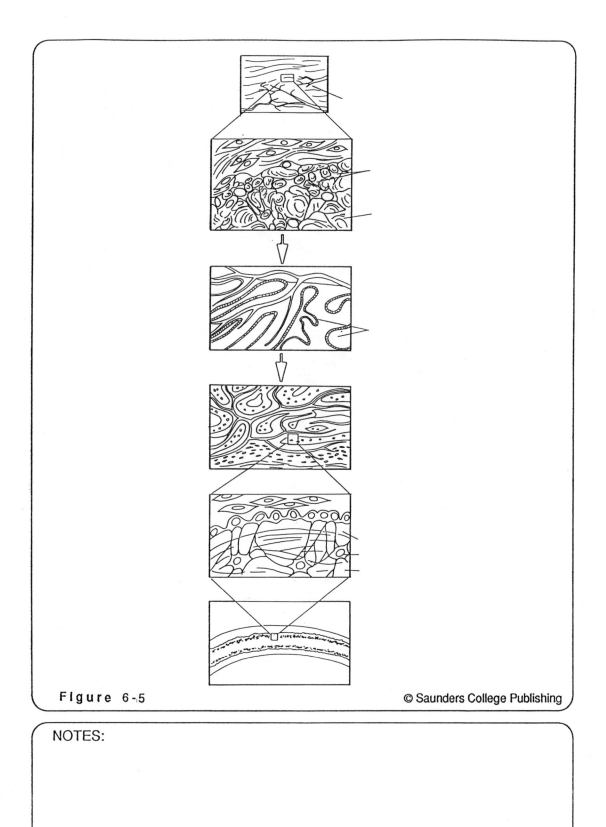

Figure 6-5

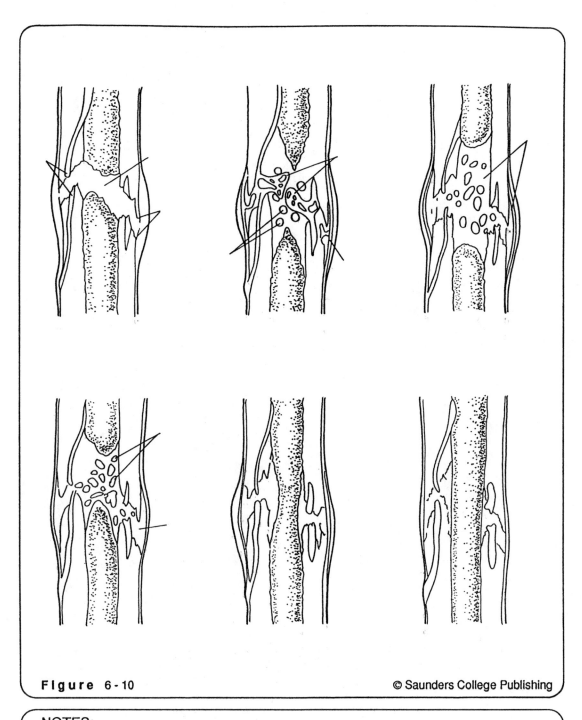

Figure 6-10

© Saunders College Publishing

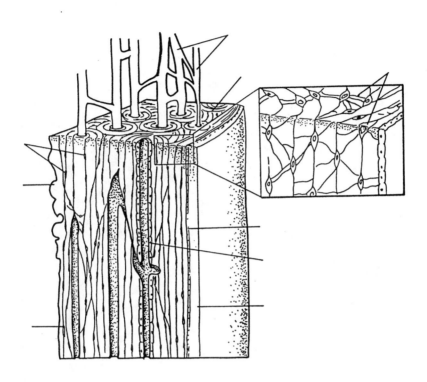

Figure 6-15

NOTES:

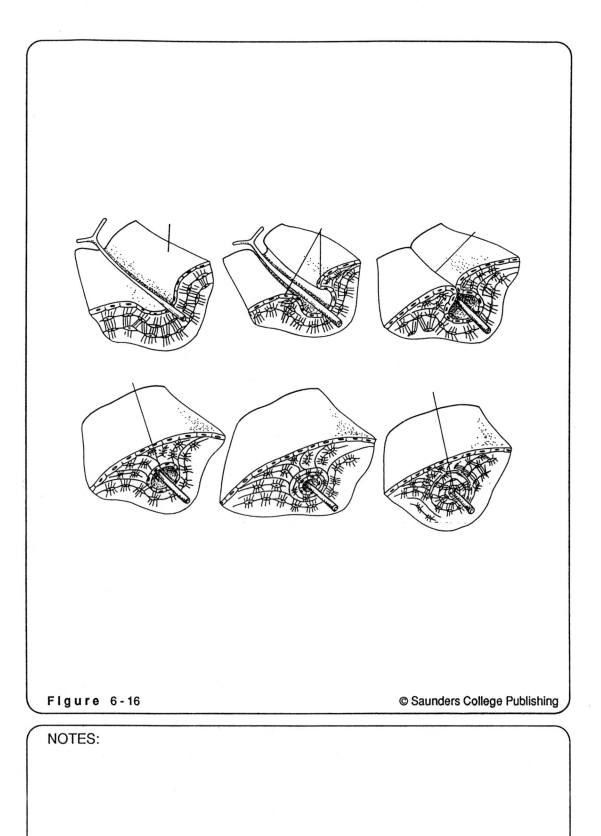

Figure 6-16

NOTES:

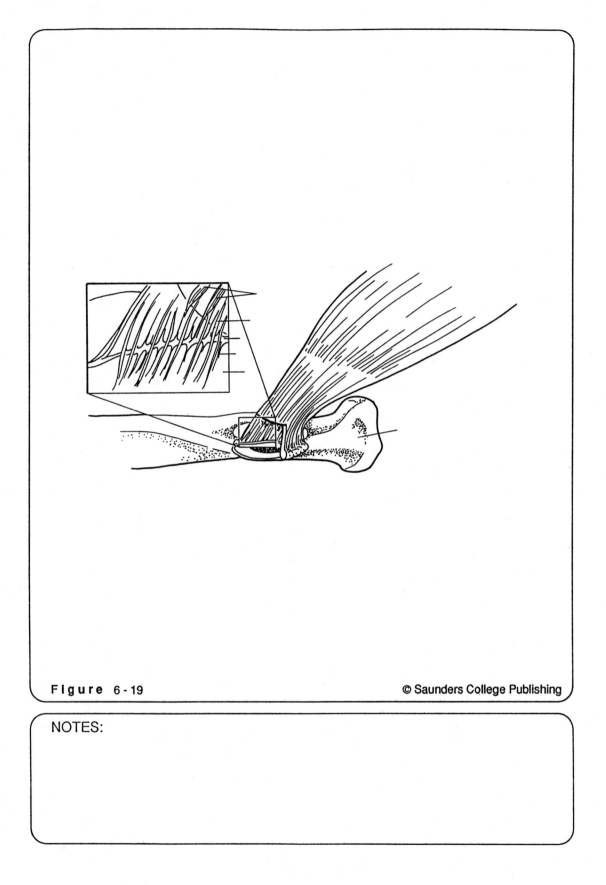

Figure 6-19

NOTES:

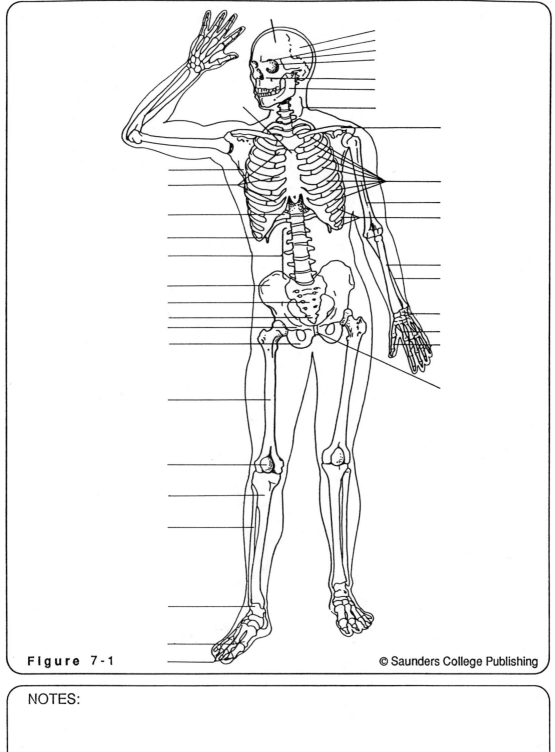

Figure 7-1

© Saunders College Publishing

NOTES:

14

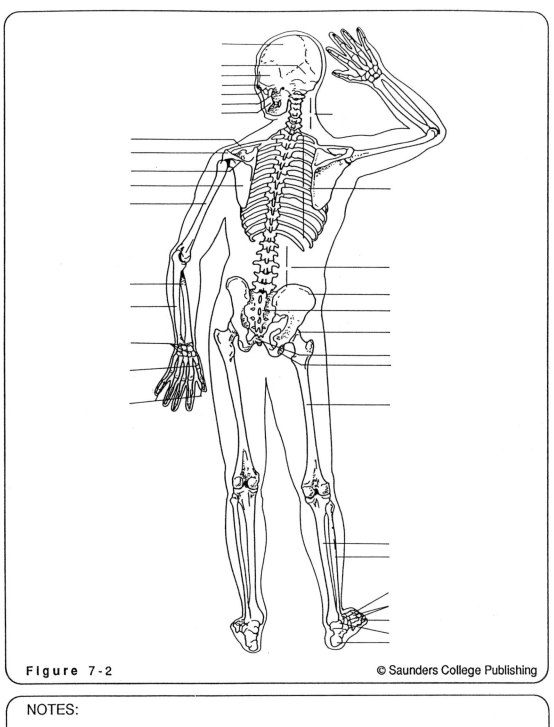

Figure 7-2

NOTES:

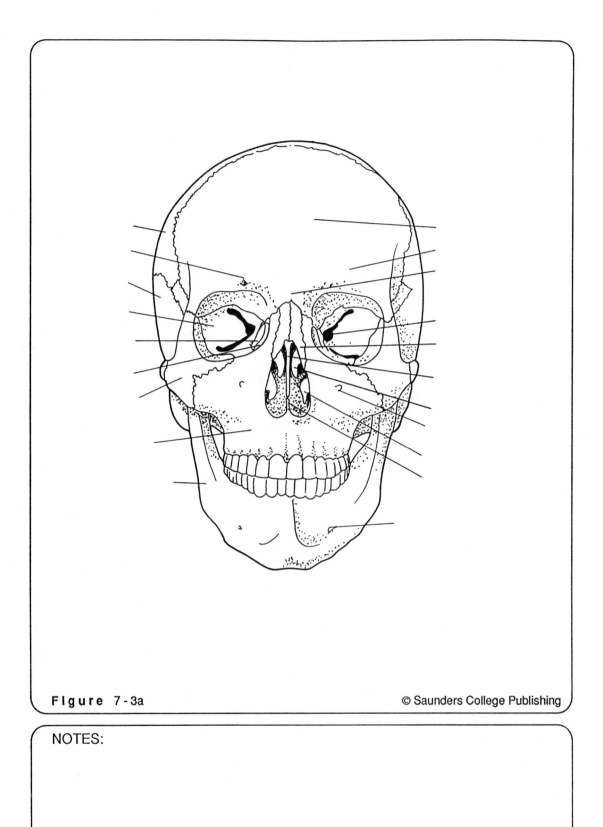

Figure 7-3a

NOTES:

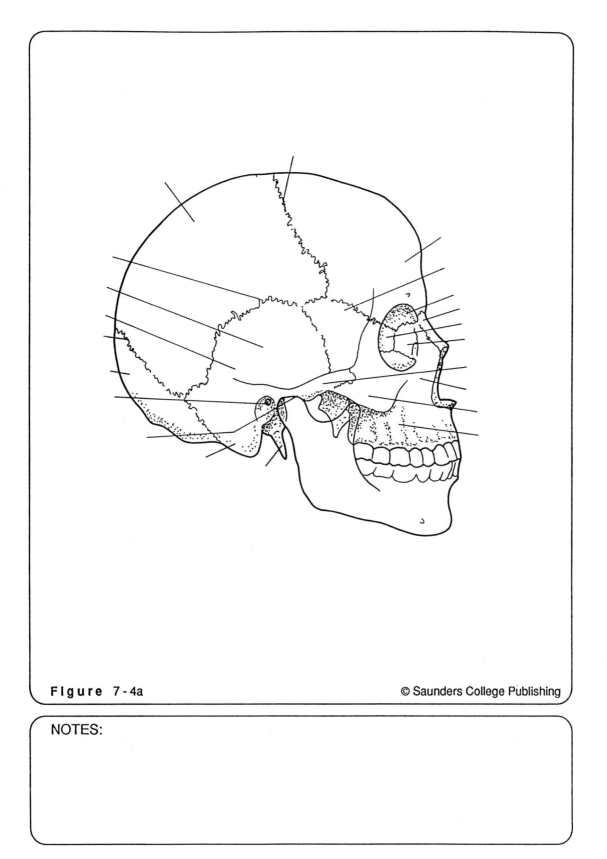

Figure 7 - 4a

NOTES:

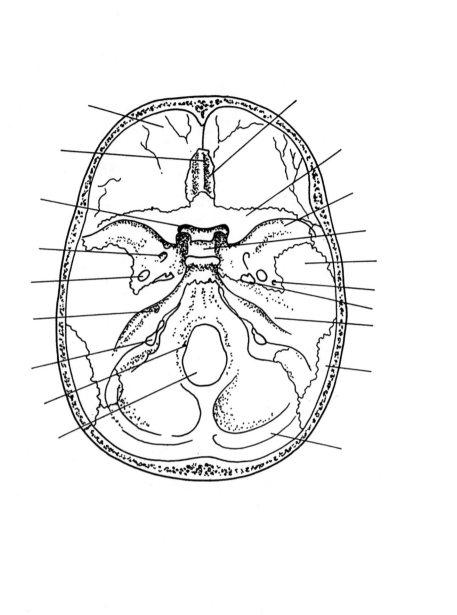

Figure 7-6a

© Saunders College Publishing

NOTES:

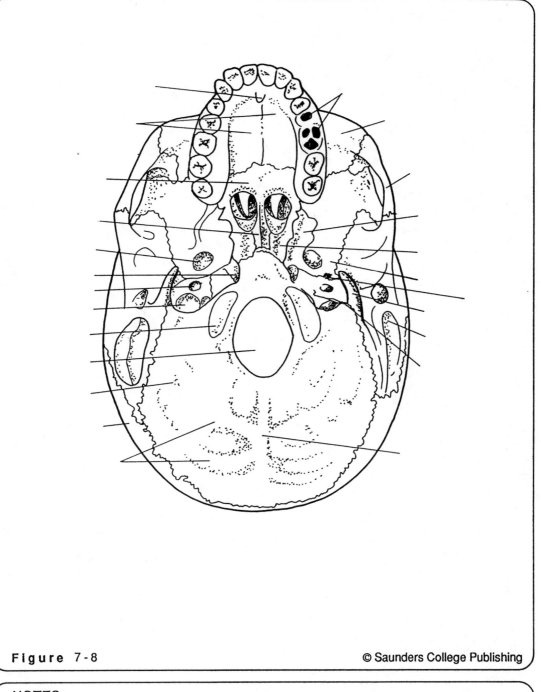

F i g u r e 7 - 8

NOTES:

Figure 18

NOTES:

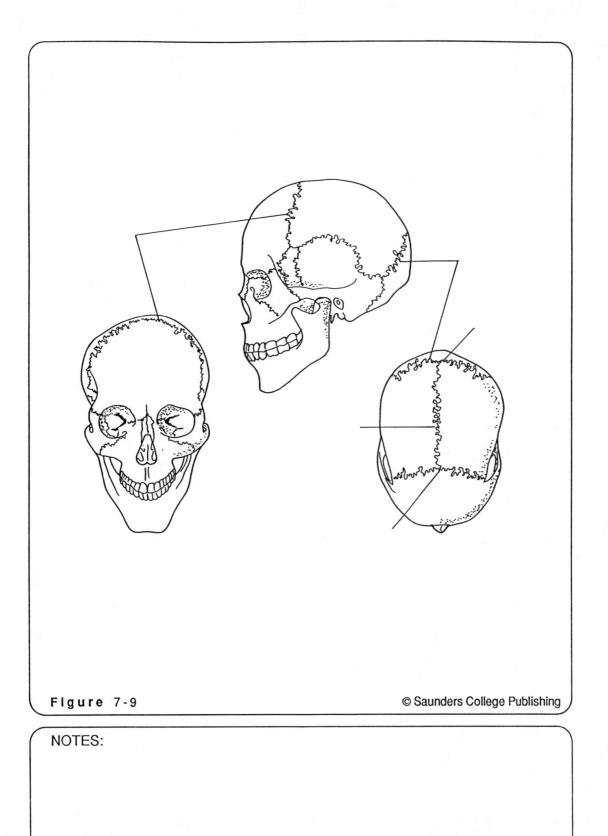

Figure 7-9

NOTES:

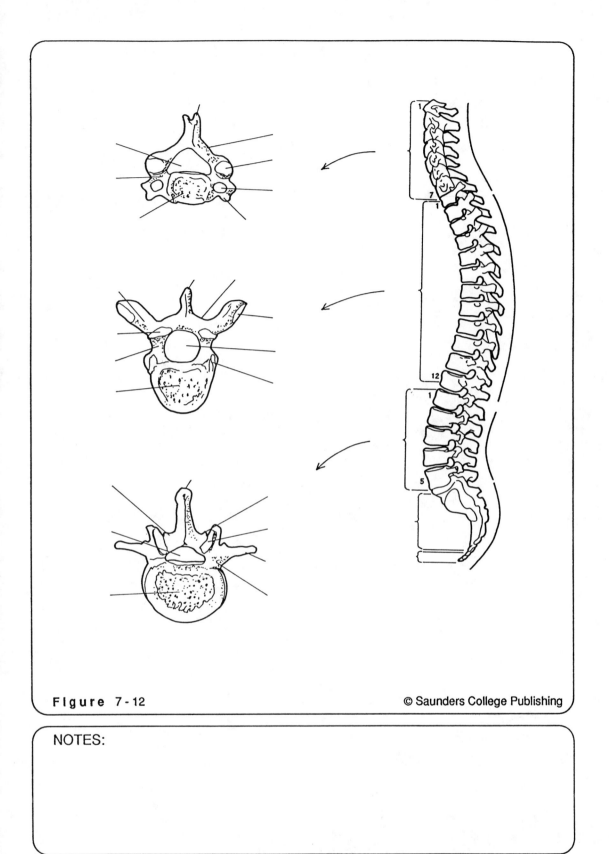

F i g u r e 7 - 12

© Saunders College Publishing

NOTES:

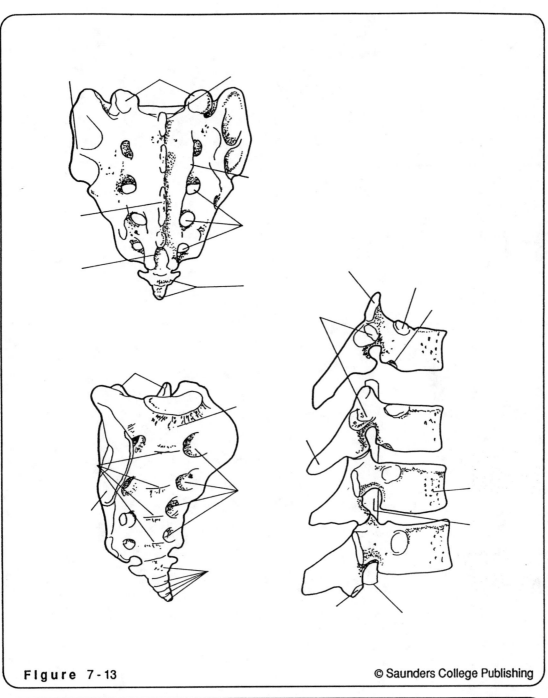

Figure 7-13

NOTES:

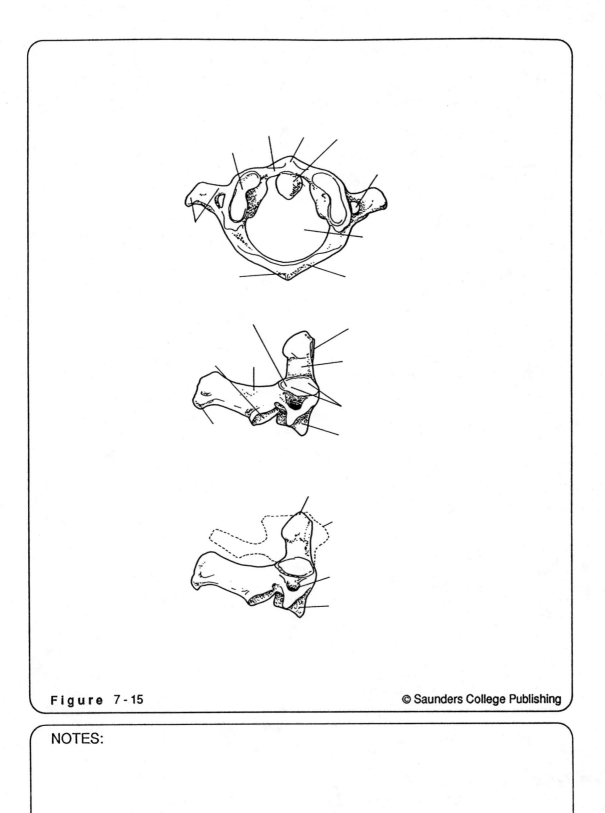

Figure 7-15

© Saunders College Publishing

NOTES:

23

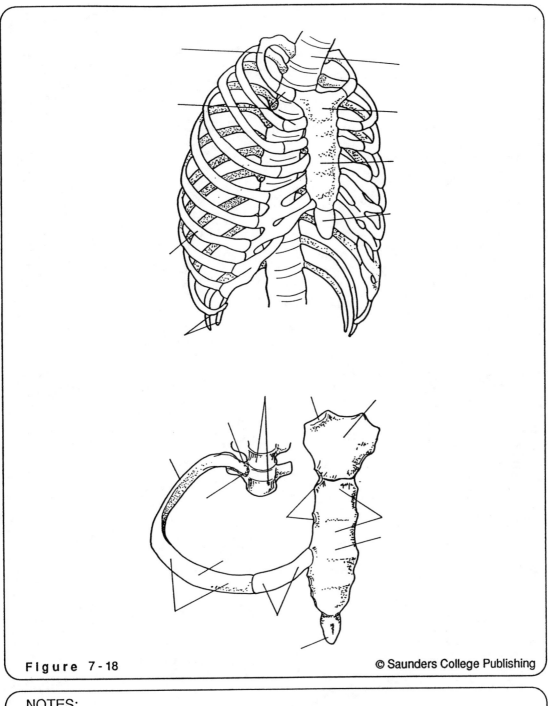

Figure 7-18

NOTES:

NOTE

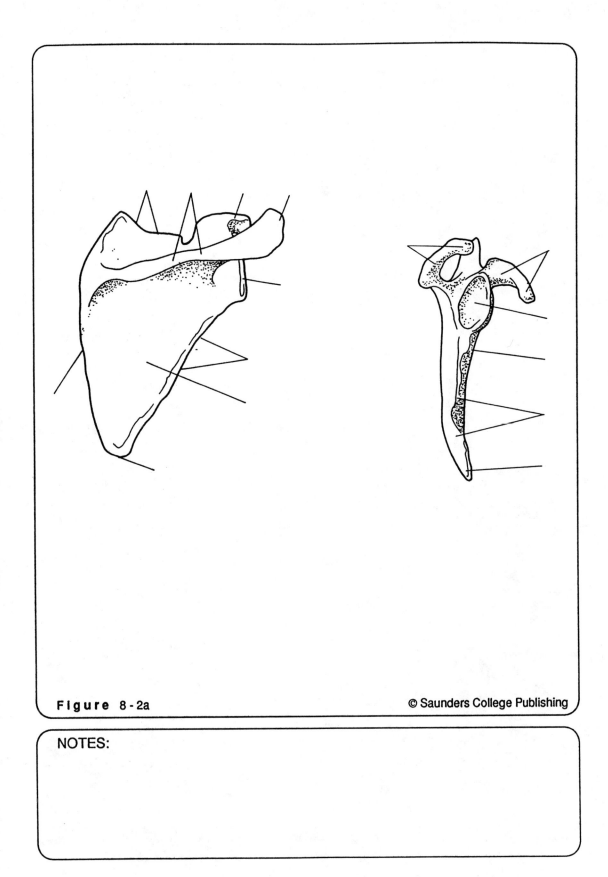

Figure 8-2a

NOTES:

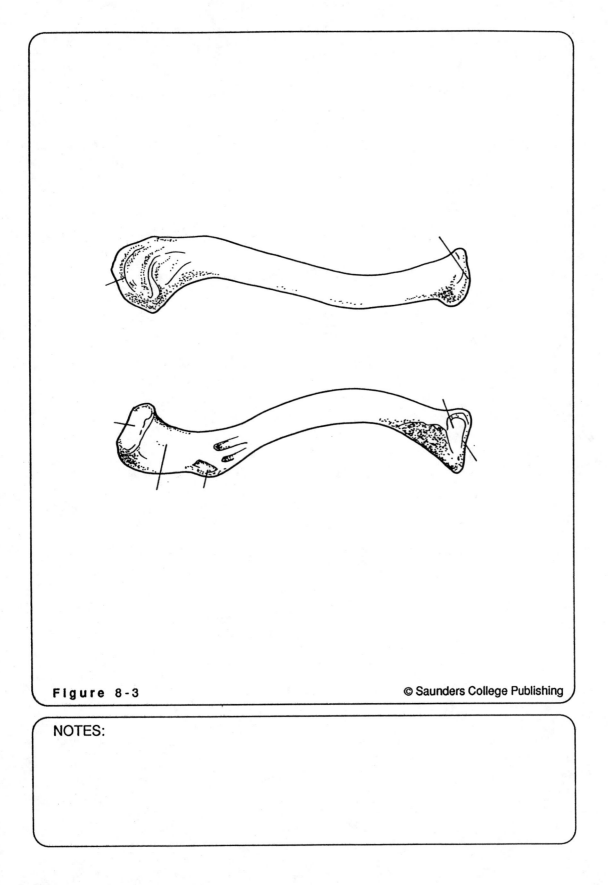

Figure 8-3

© Saunders College Publishing

NOTES:

26

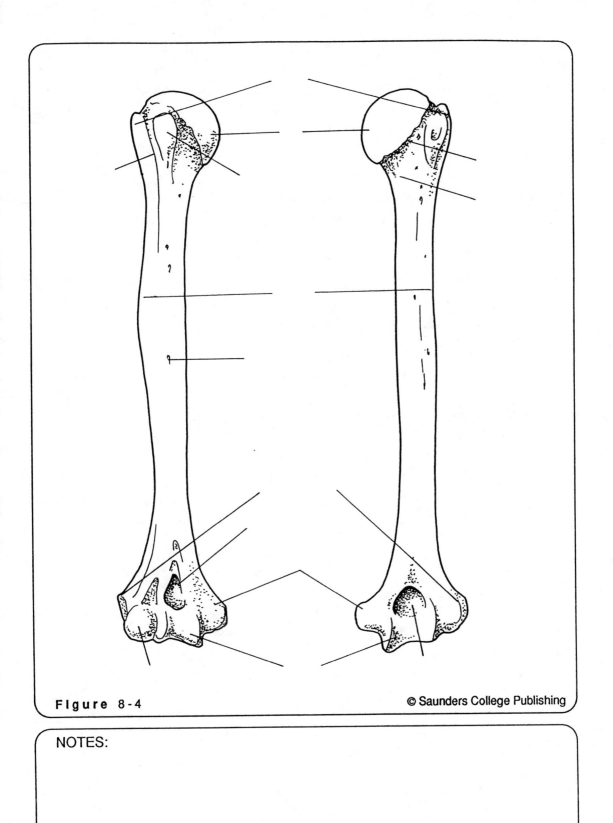

Figure 8-4

© Saunders College Publishing

NOTES:

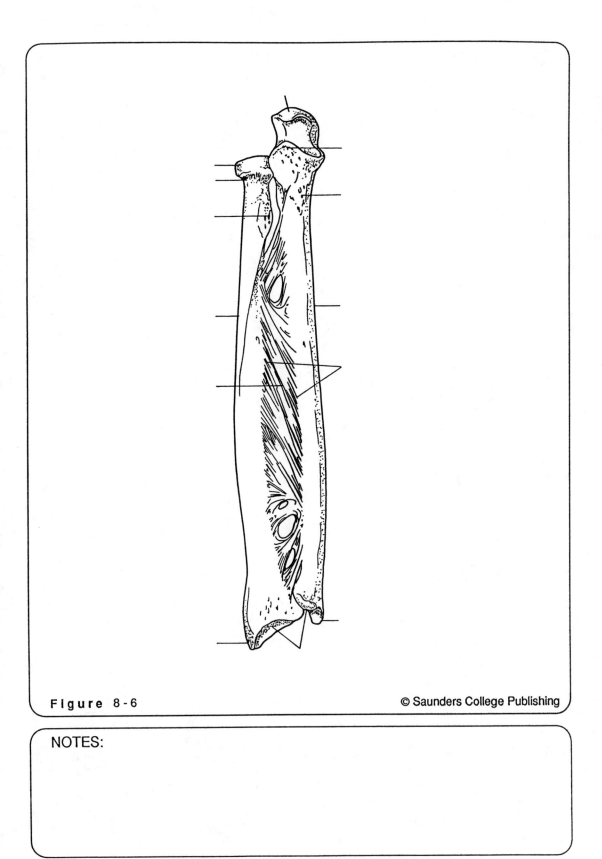

Figure 8-6

NOTES:

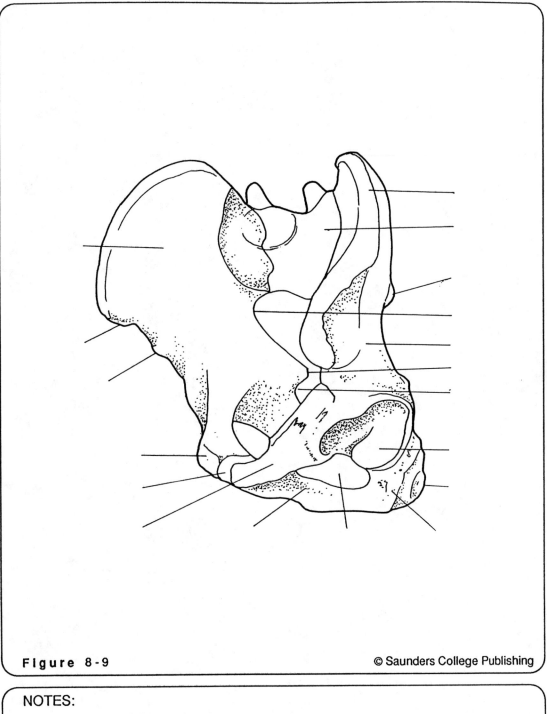

Figure 8-9

NOTES:

Figure 3-6

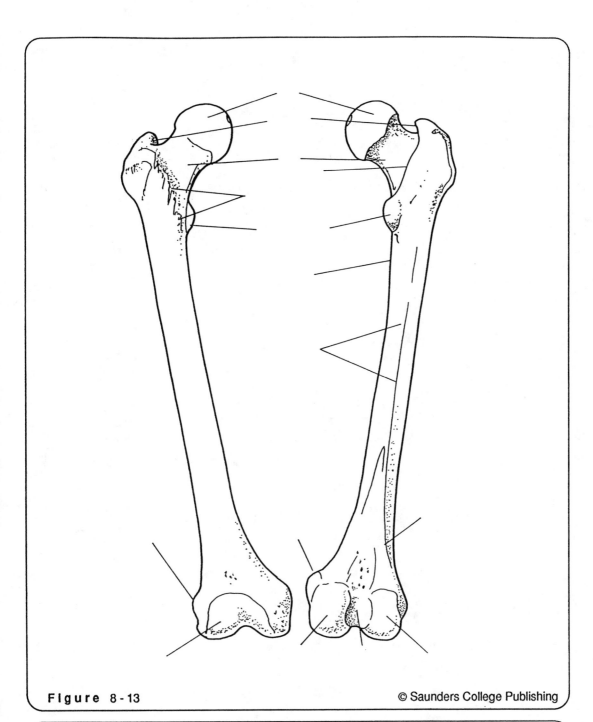

Figure 8-13

NOTES:

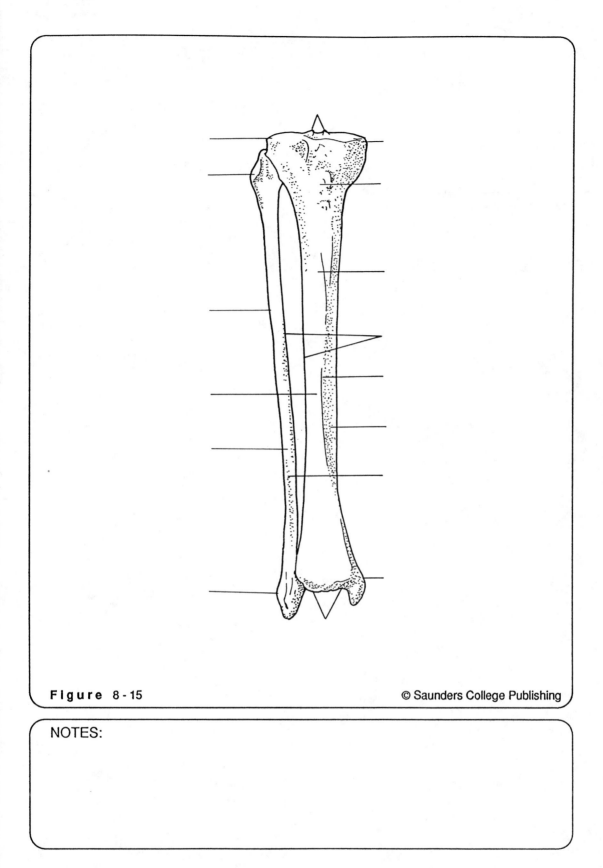

F i g u r e 8 - 15

NOTES:

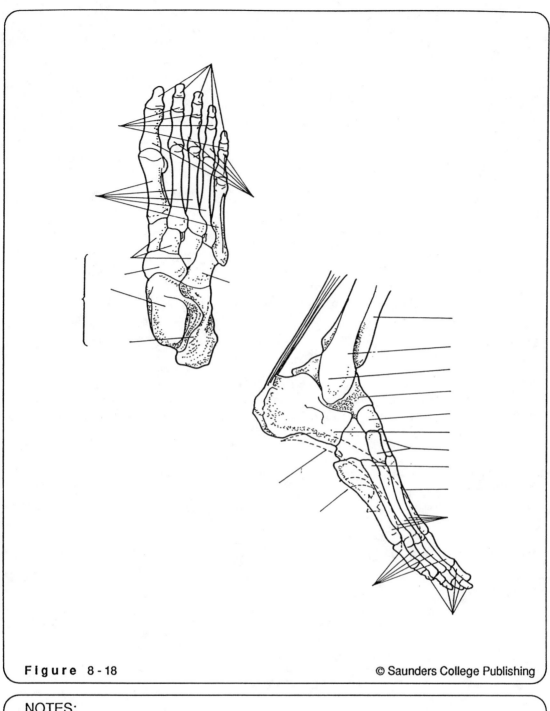

Figure 8-18

NOTES:

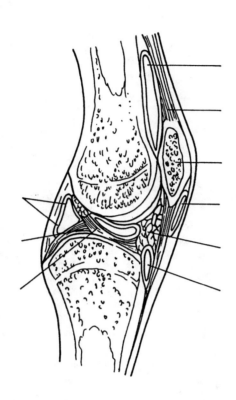

Figure 9-1

NOTES:

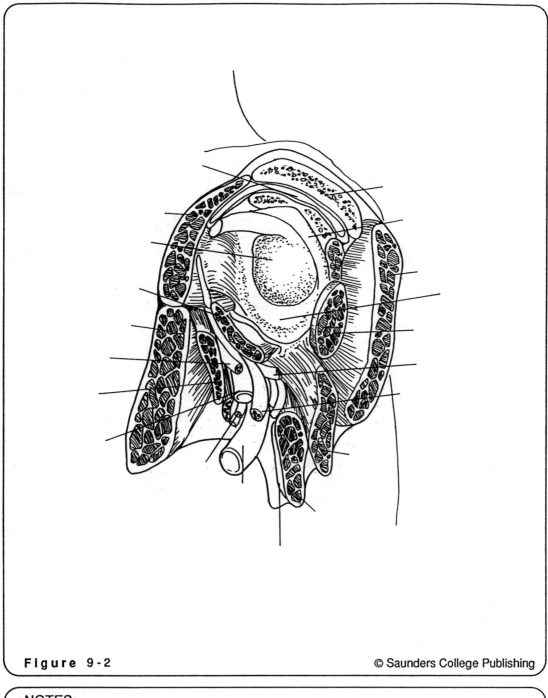

Figure 9-2

© Saunders College Publishing

NOTES:

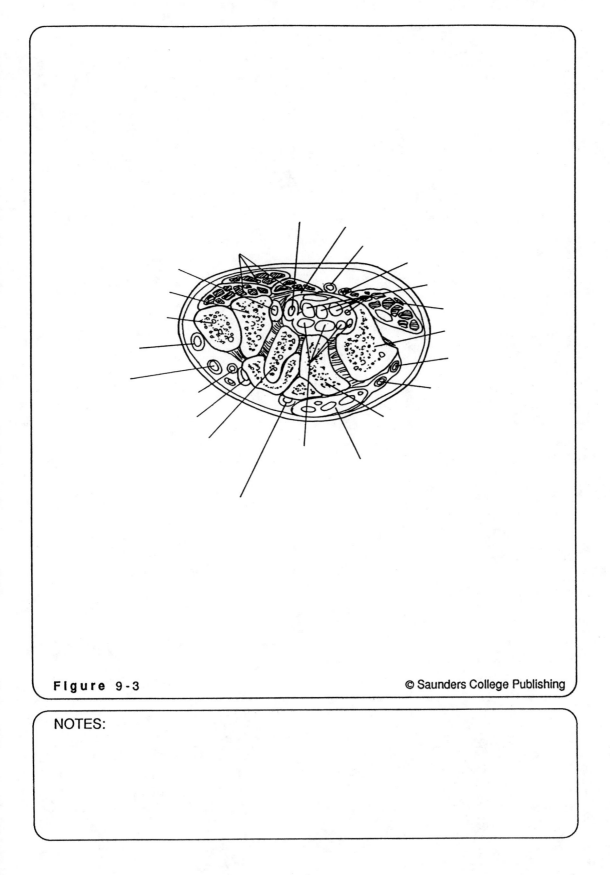

Figure 9-3

NOTES:

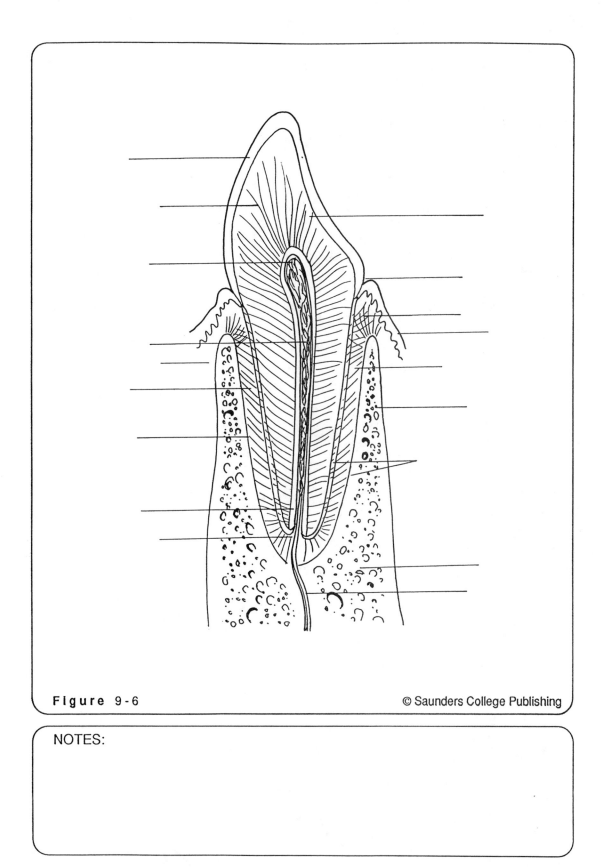

Figure 9-6

NOTES:

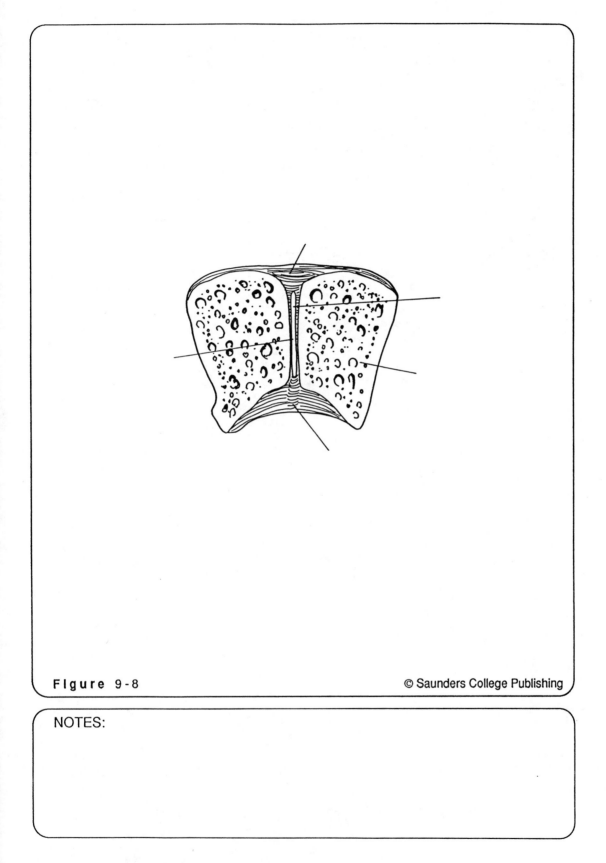

Figure 9-8

NOTES:

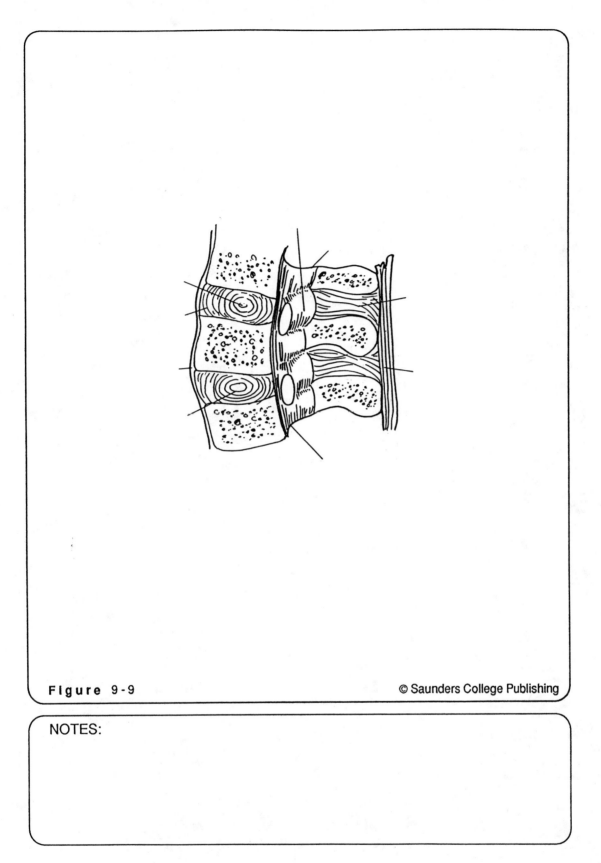

Figure 9-9

NOTES:

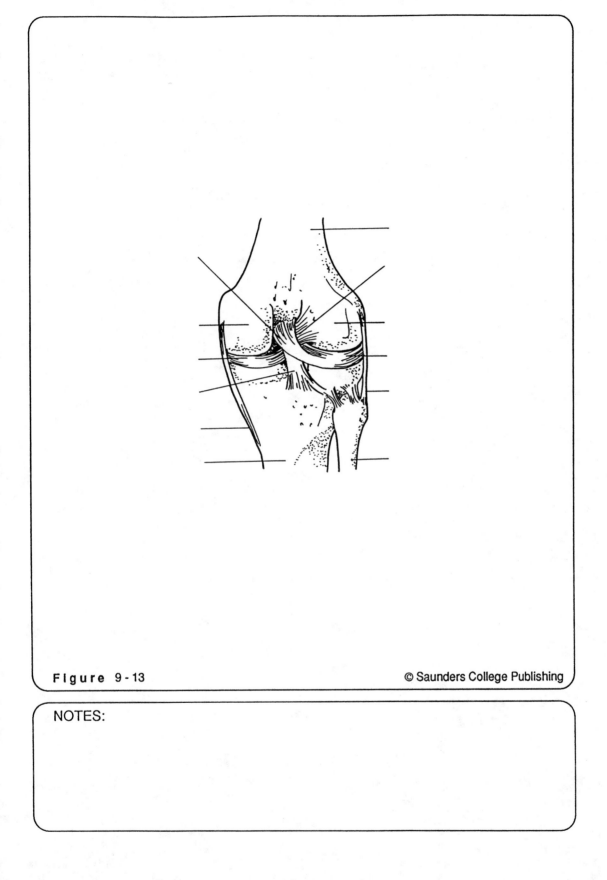

Figure 9-13

NOTES:

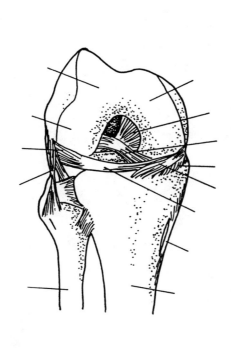

Figure 9-16

NOTES:

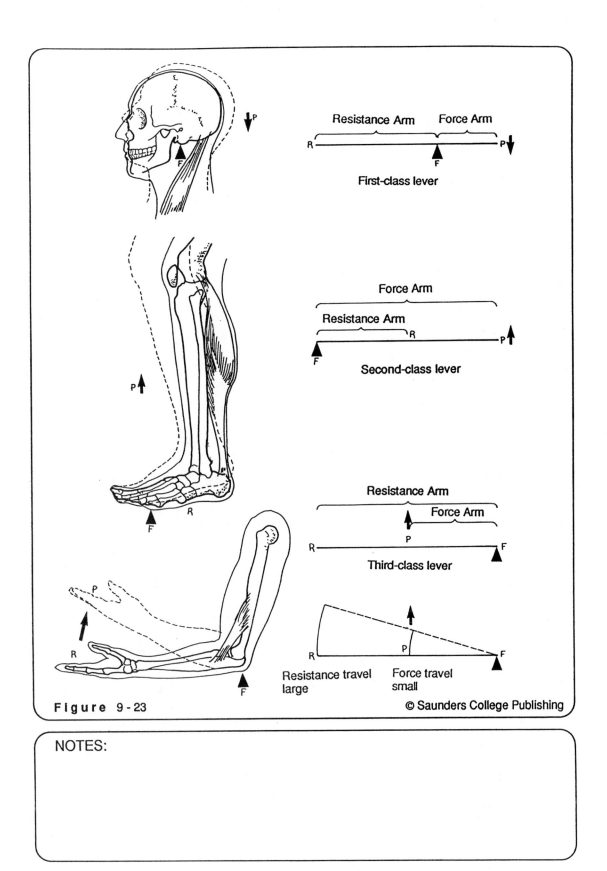

Resistance Arm   Force Arm

R ———————————————— F ———— P↓

First-class lever

Force Arm

Resistance Arm

F
Second-class lever

Resistance Arm

Force Arm

R ———————————————— F

Third-class lever

Resistance travel   Force travel
large   small

Figure 9-23   © Saunders College Publishing

NOTES:

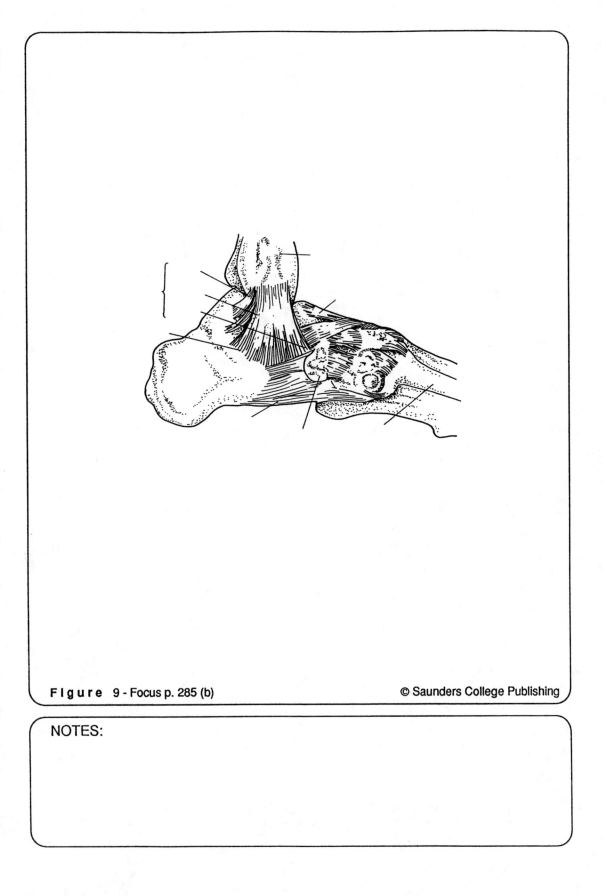

F i g u r e 9 - Focus p. 285 (b)

NOTES:

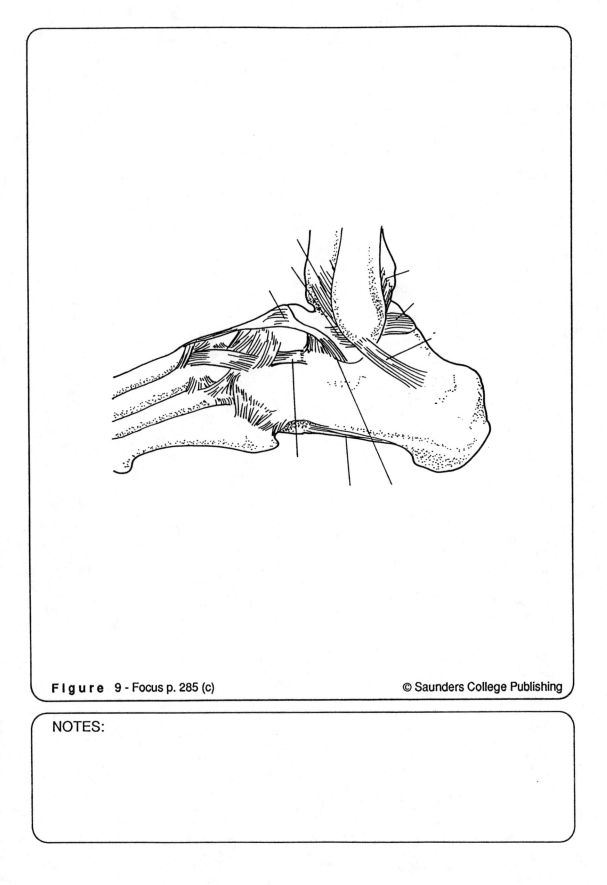

F i g u r e 9 - Focus p. 285 (c)   © Saunders College Publishing

NOTES:

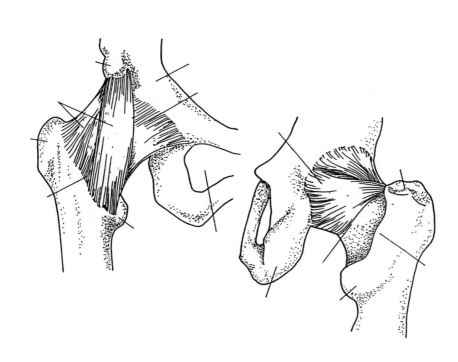

NOTES:

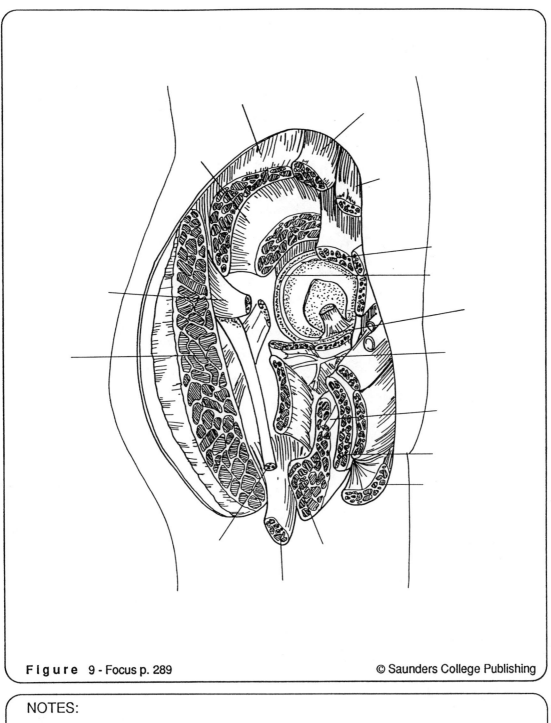

Figure 9 - Focus p. 289

NOTES:

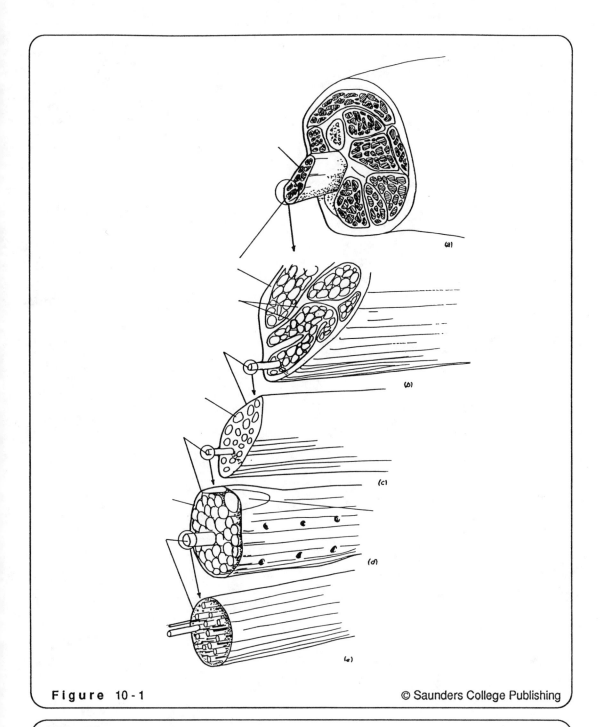

F i g u r e 10 - 1

NOTES:

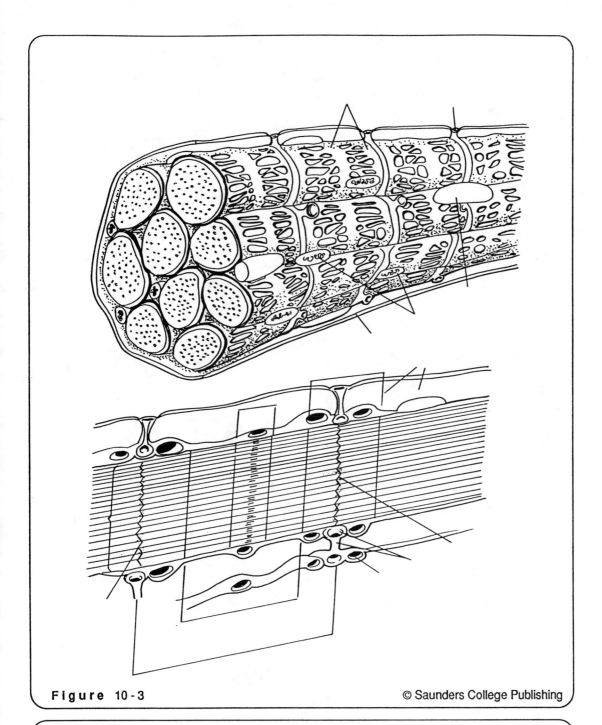

Figure 10-3

NOTES:

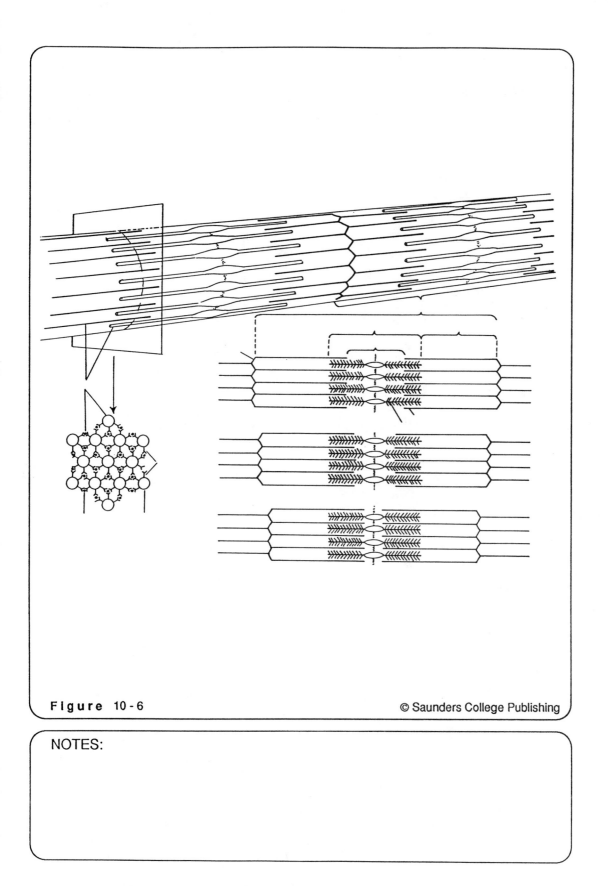

Figure 10-6

NOTES:

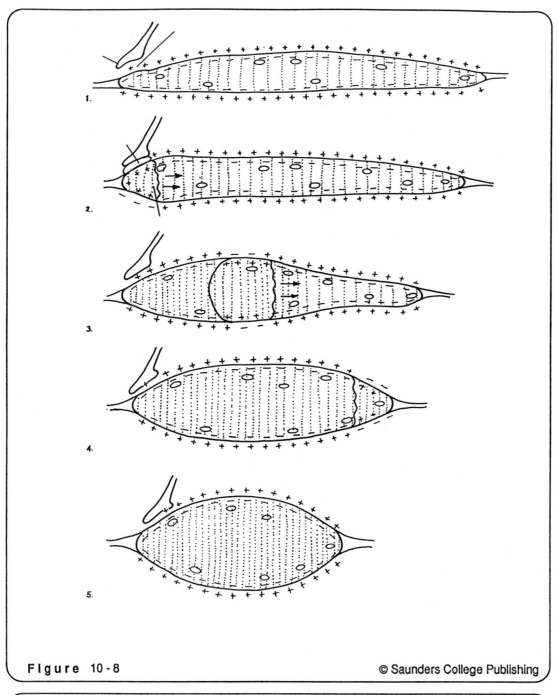

Figure 10-8

© Saunders College Publishing

NOTES:

49

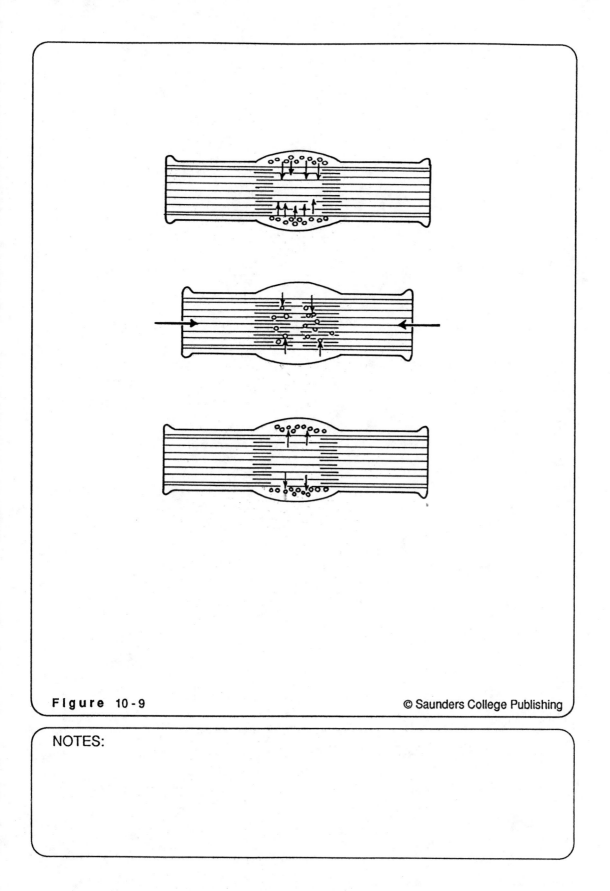

Figure 10-9

NOTES:

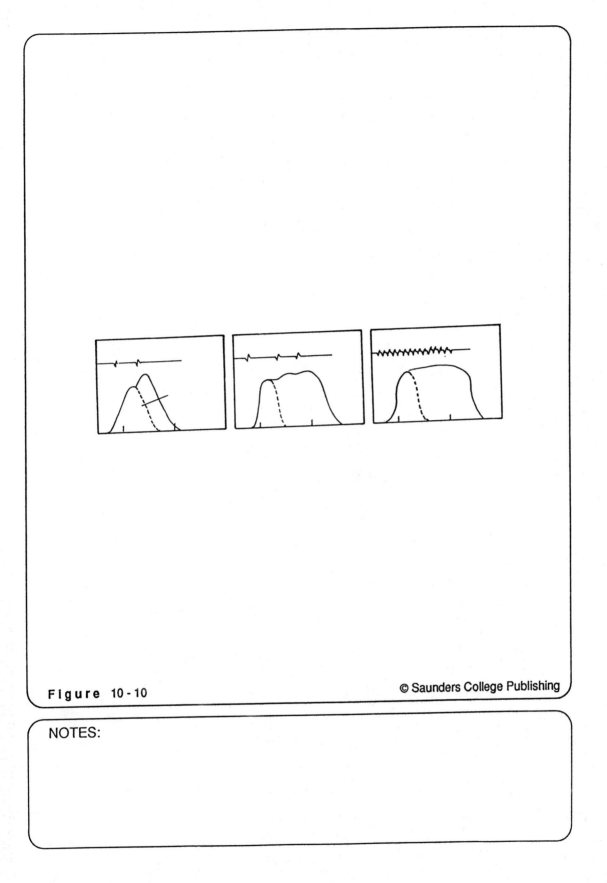

F i g u r e 10 - 10

NOTES:

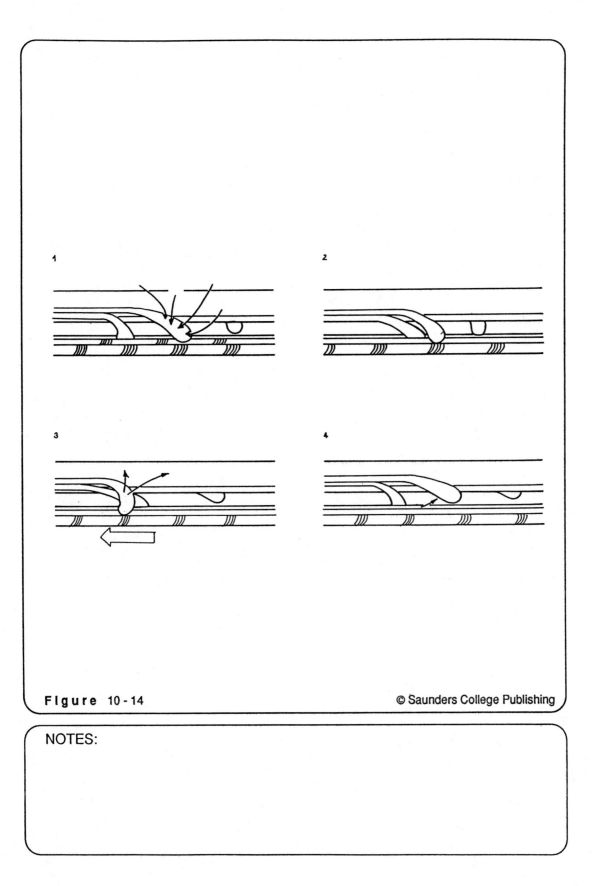

Figure 10 - 14

© Saunders College Publishing

NOTES:

Figure 10-14

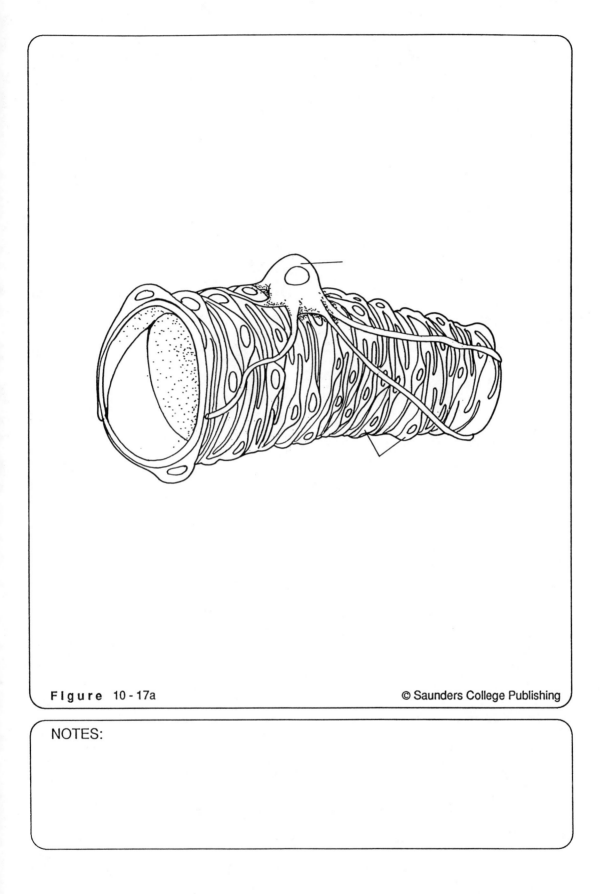

Figure 10-17a

NOTES:

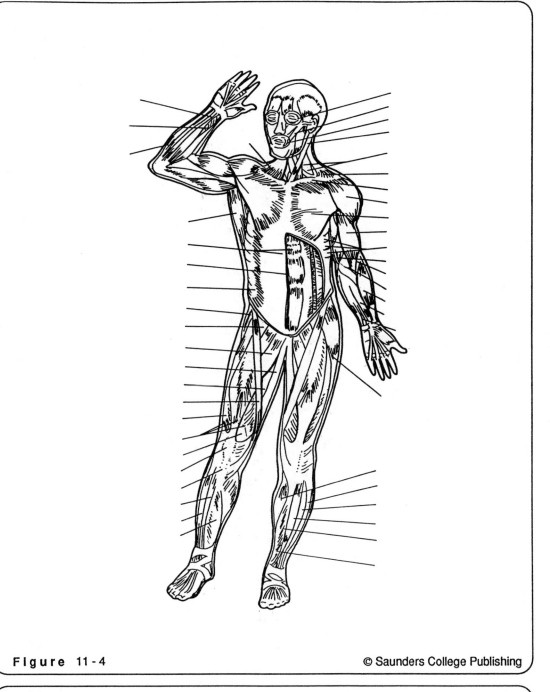

Figure 11-4

NOTES:

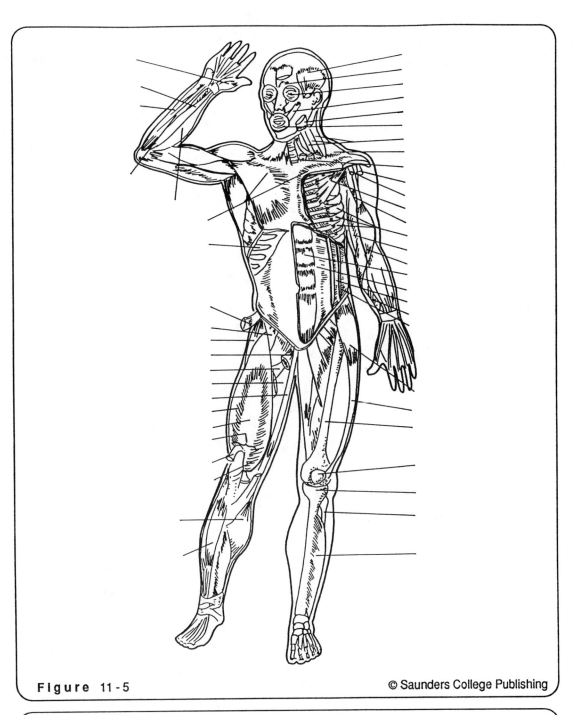

Figure 11-5

NOTES:

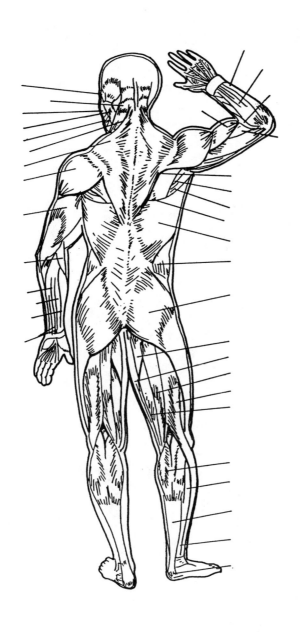

Figure 11-6

NOTES:

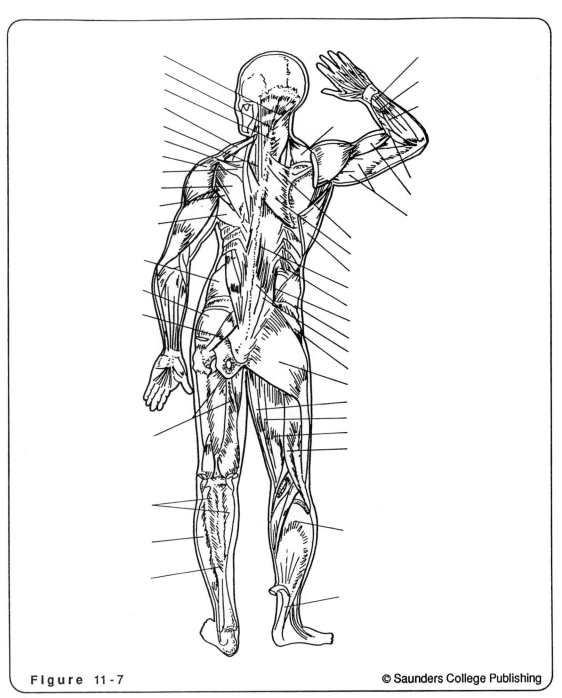

Figure 11-7

© Saunders College Publishing

NOTES:

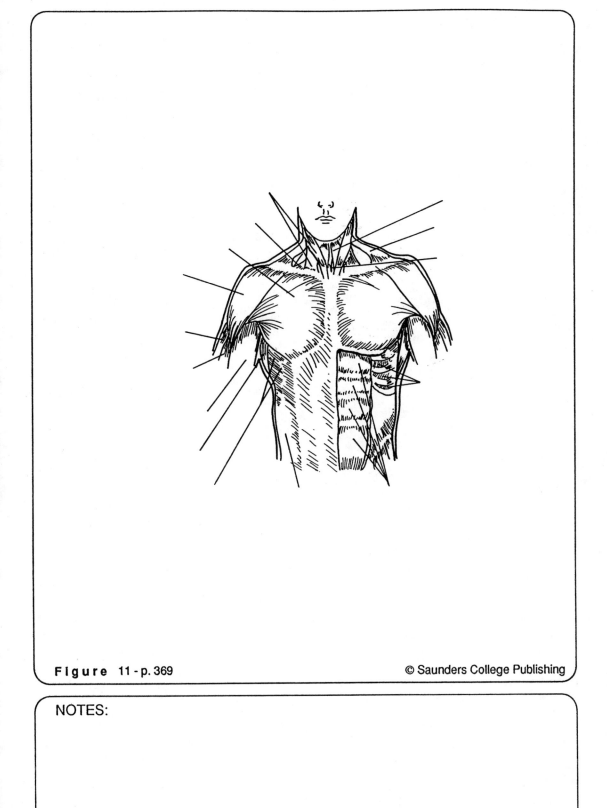

F i g u r e 11 - p. 369

NOTES:

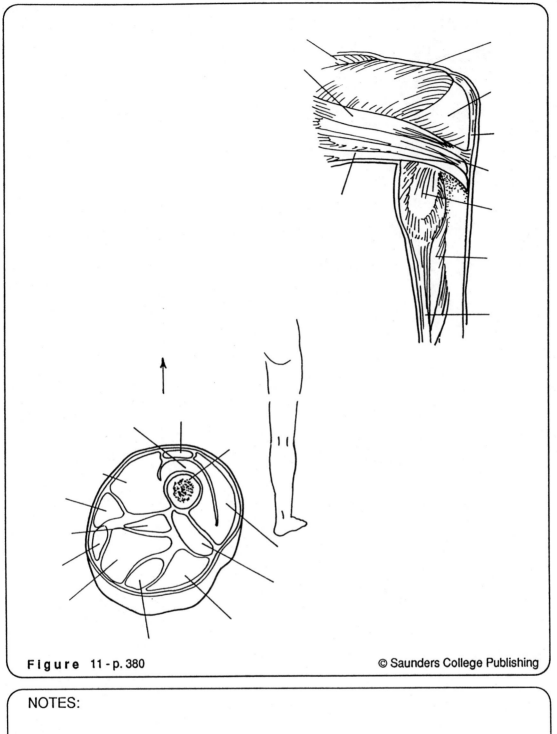

Figure 11 - p. 380

NOTES:

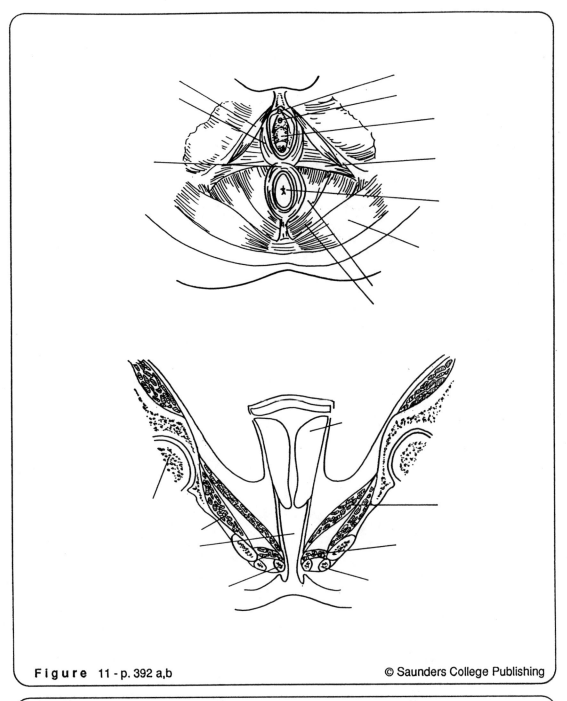

F i g u r e 11 - p. 392 a,b

NOTES:

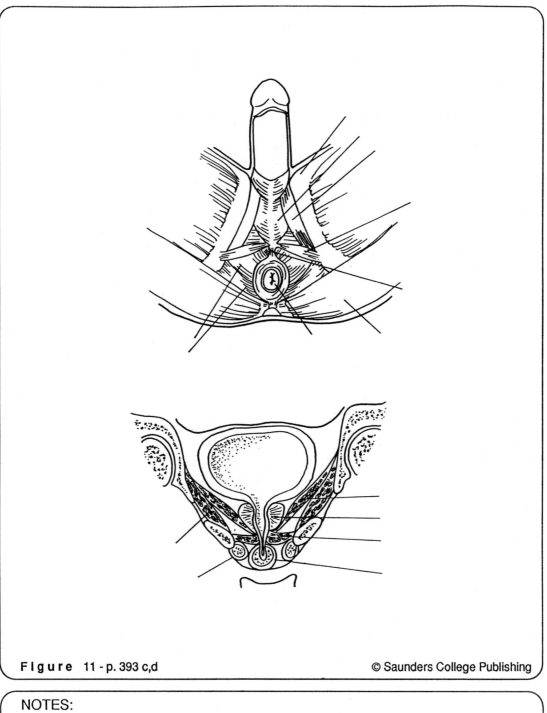

Figure 11 - p. 393 c,d

NOTES:

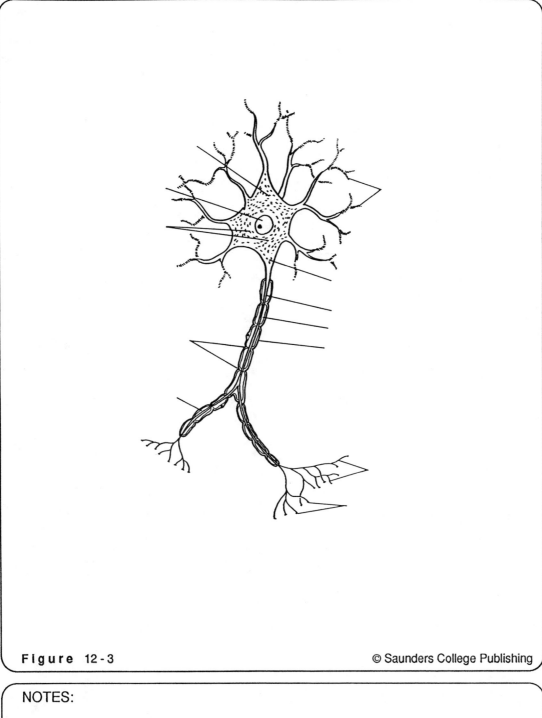

Figure 12-3

© Saunders College Publishing

NOTES:

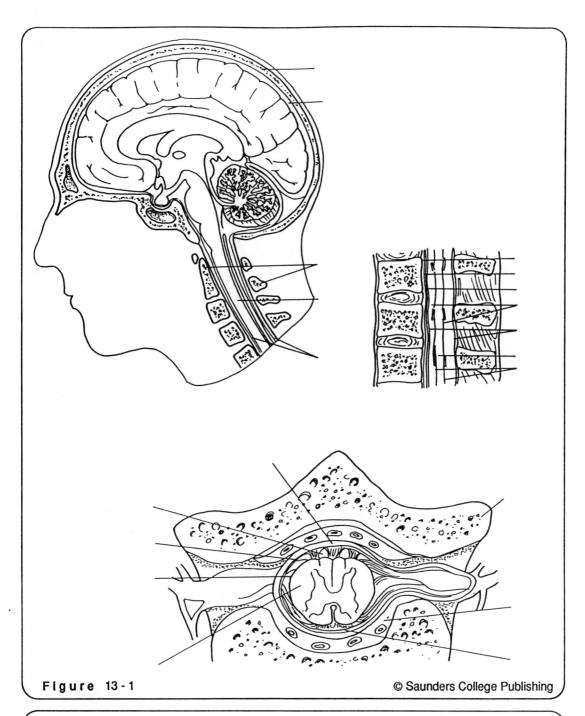

Figure 13-1

NOTES:

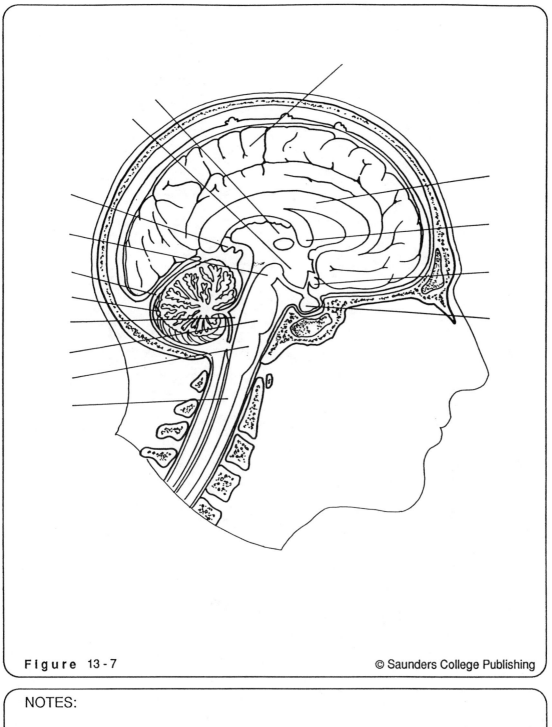

Figure 13-7

NOTES:

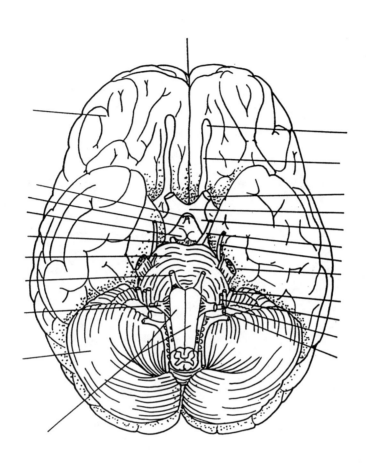

Figure 13 - 9

© Saunders College Publishing

NOTES:

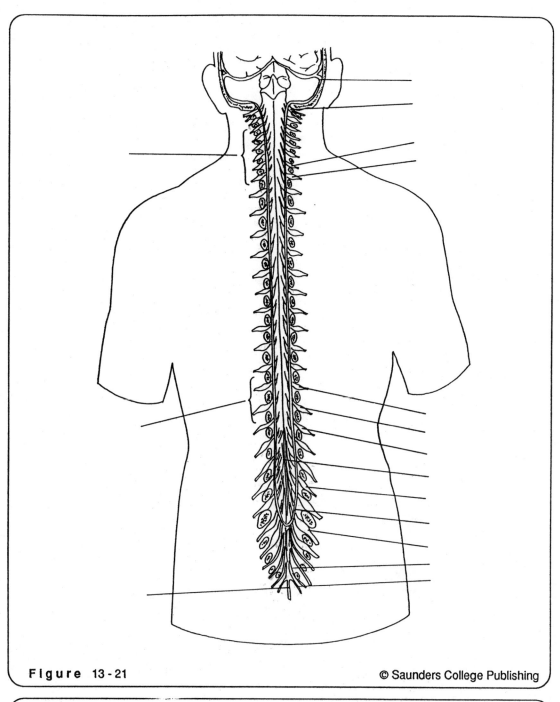

Figure 13-21   © Saunders College Publishing

NOTES:

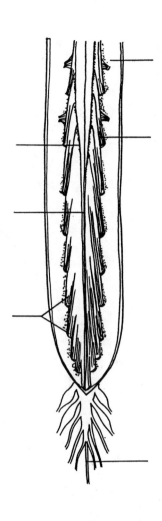

Figure 13-23

NOTES:

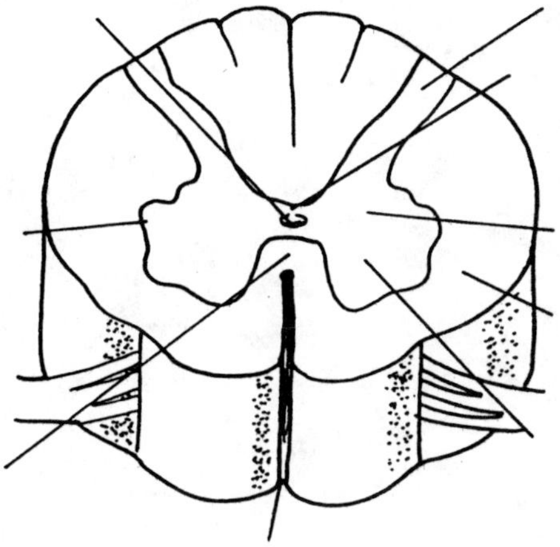

Figure 13-24

© Saunders College Publishing

NOTES:

Figure 13-26

NOTES

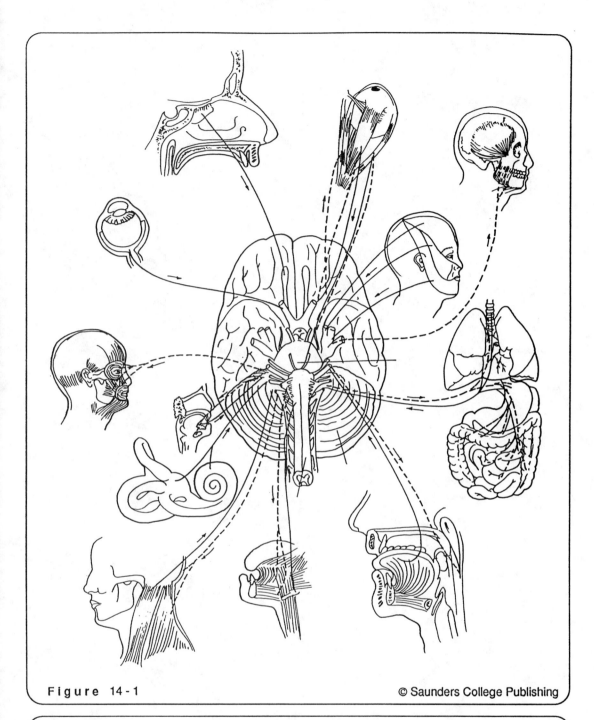

Figure 14-1

NOTES:

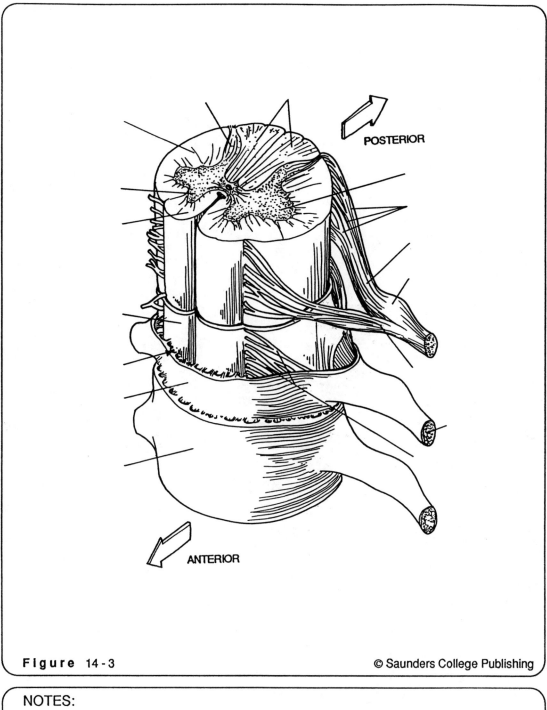

POSTERIOR

ANTERIOR

Figure 14-3

© Saunders College Publishing

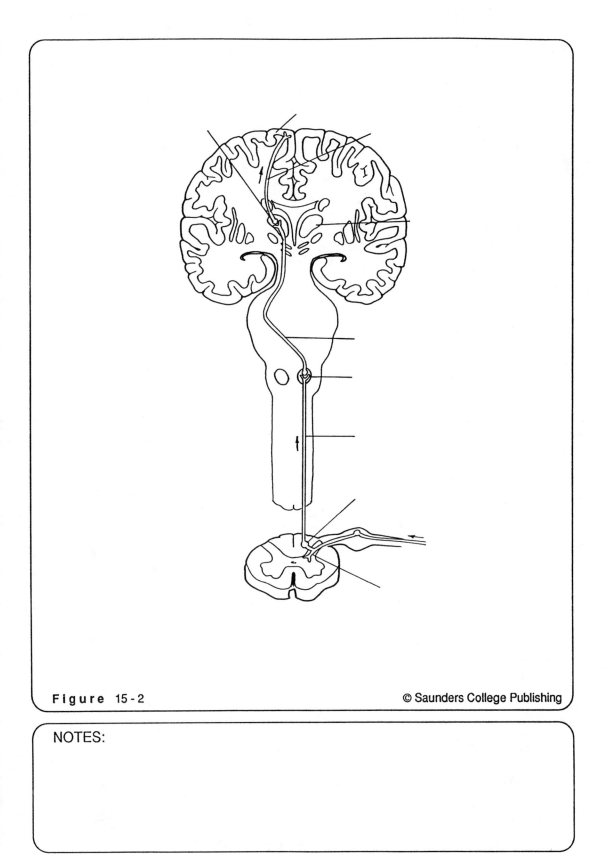

Figure 15-2

NOTES:

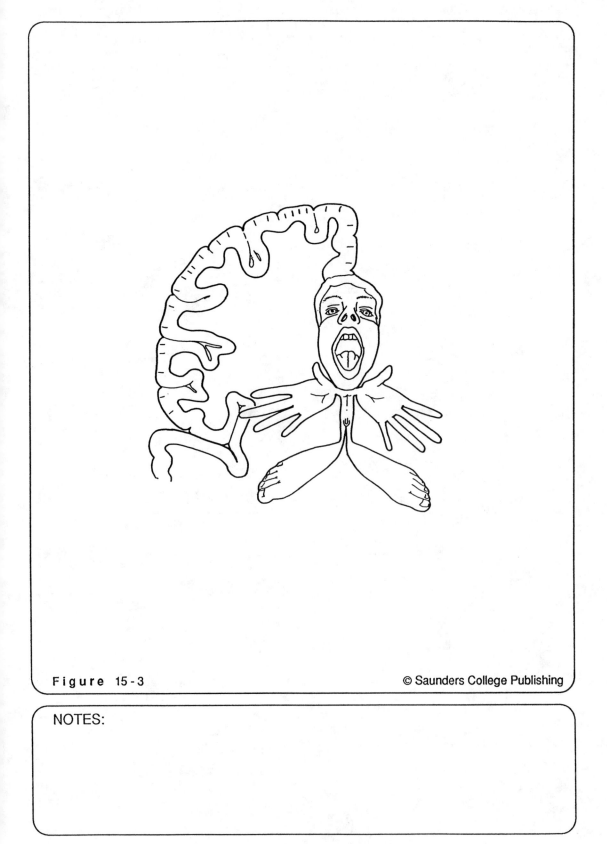

Figure 15-3

© Saunders College Publishing

NOTES:

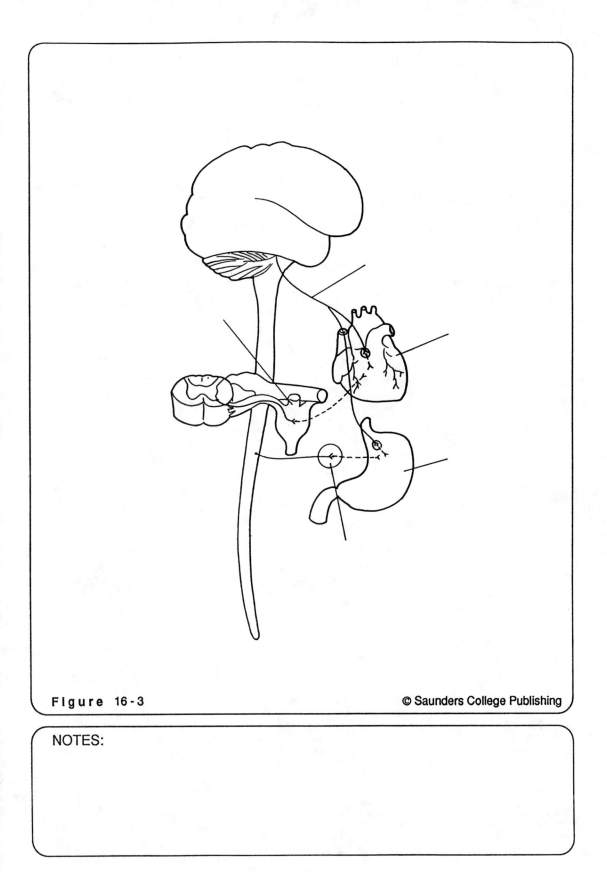

Figure 16-3

NOTES:

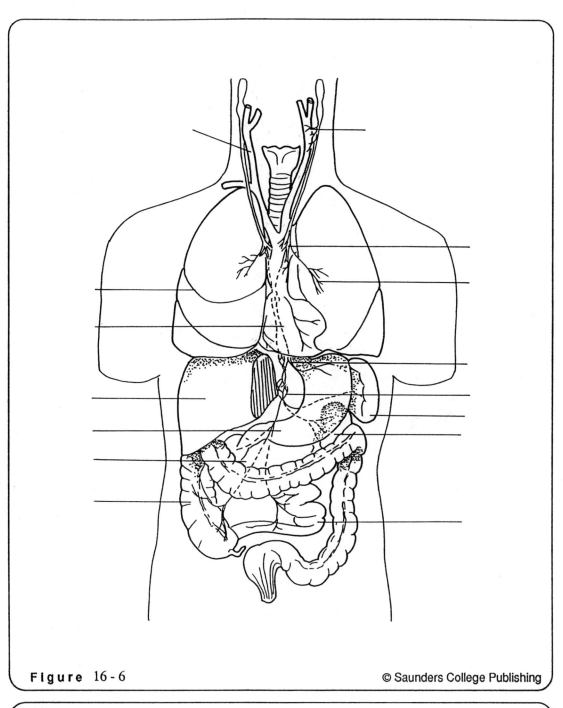

Figure 16-6

© Saunders College Publishing

NOTES:

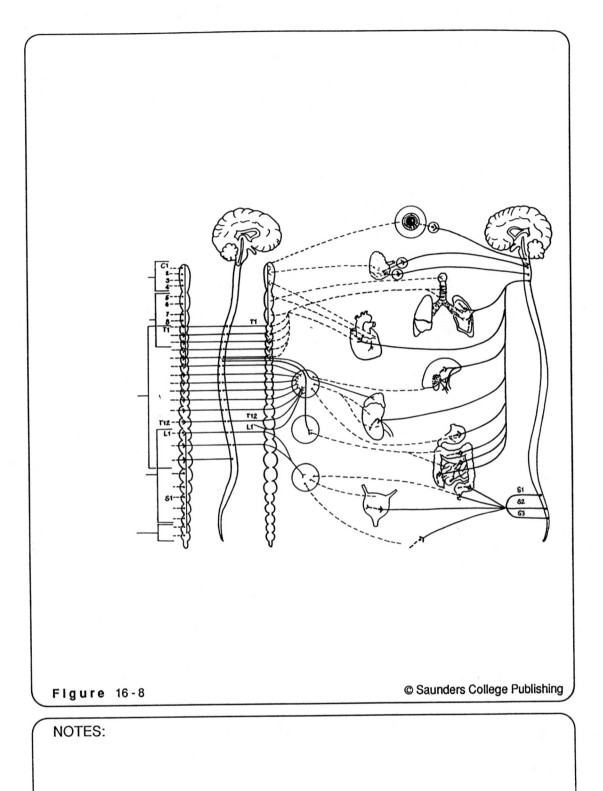

Figure 16-8

NOTES:

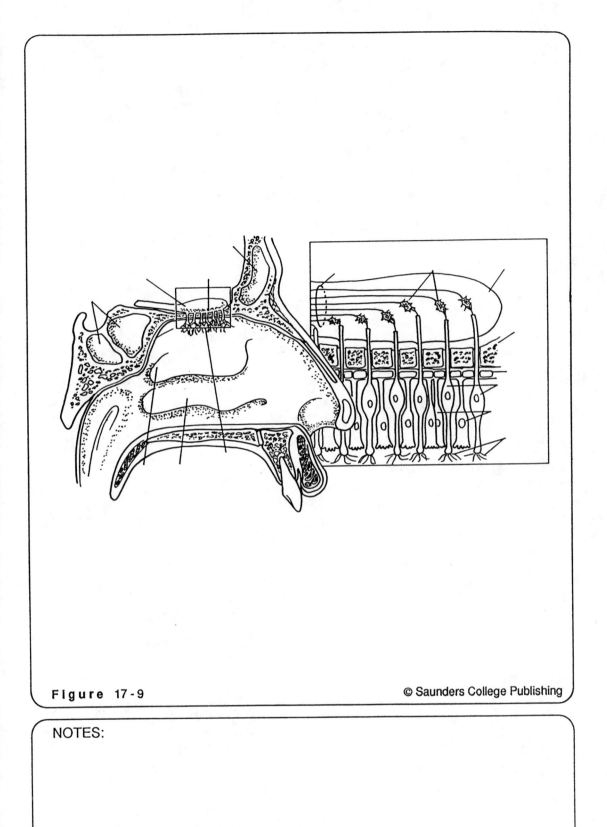

Figure 17-9

NOTES:

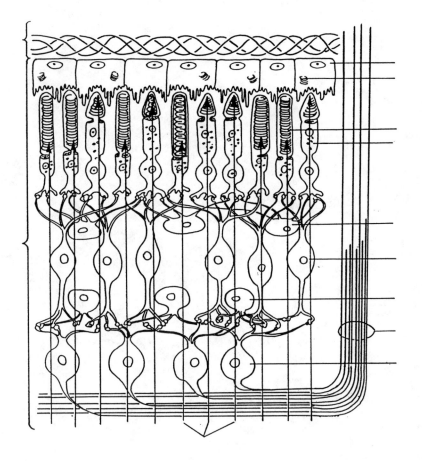

Figure 17 - 15a

© Saunders College Publishing

NOTES:

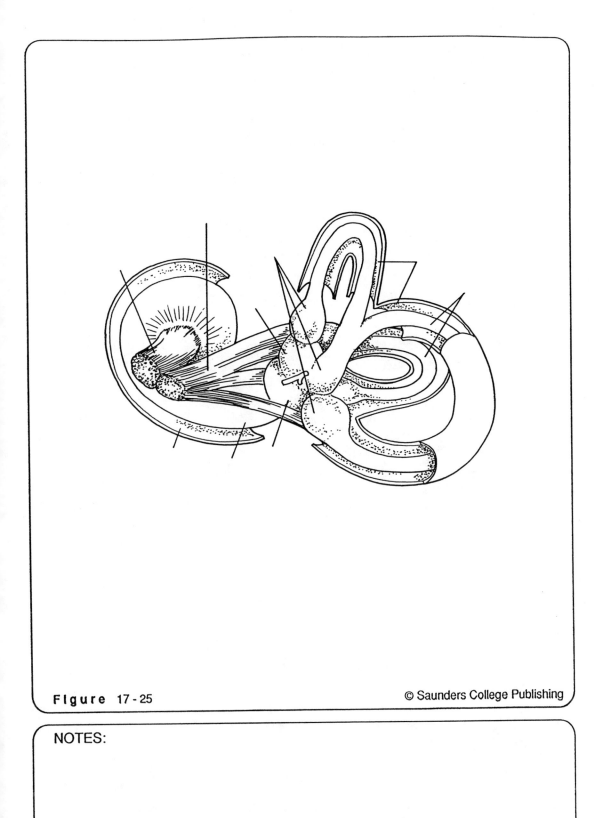

Figure 17-25

NOTES:

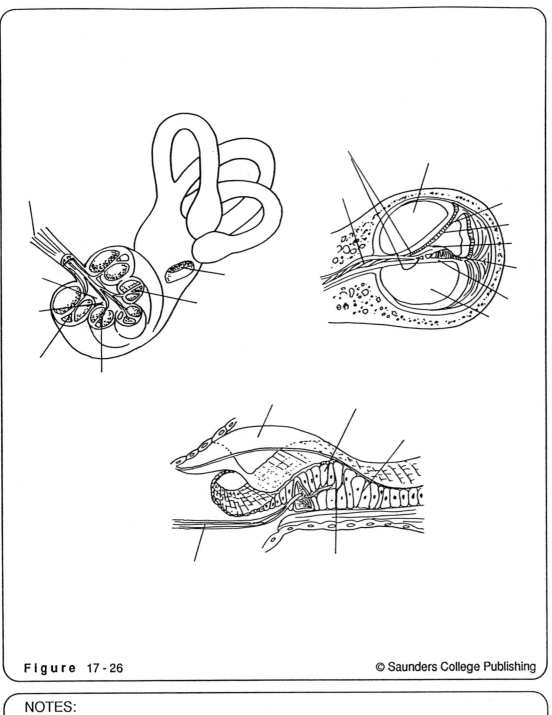

Figure 17-26

NOTES:

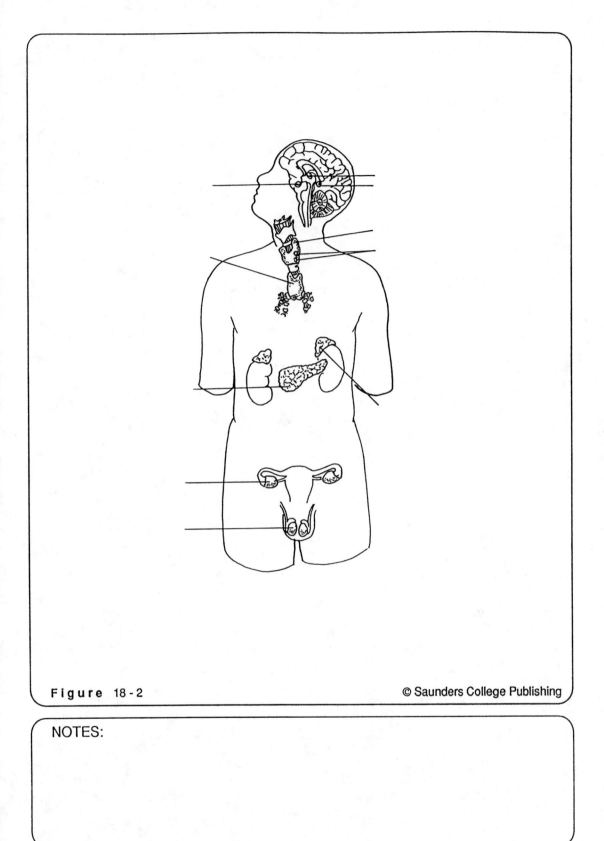

Figure 18-2

© Saunders College Publishing

NOTES:

Figure 18-7

NOTES

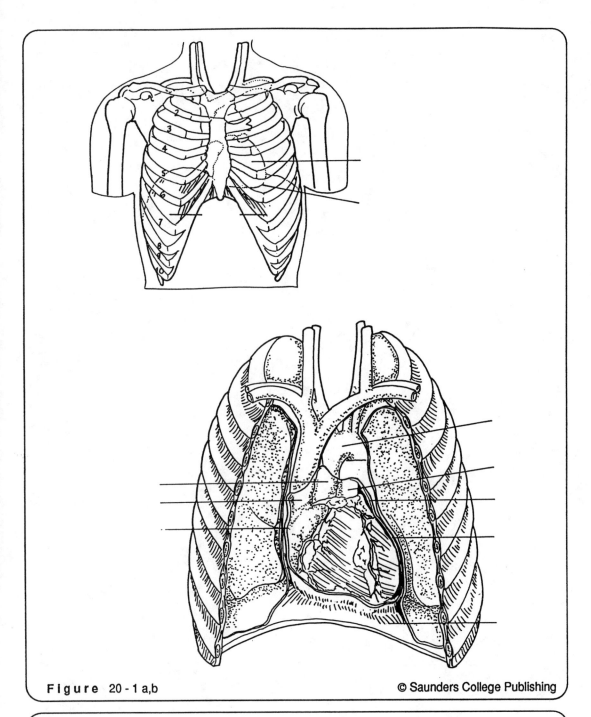

Figure 20 - 1 a,b

NOTES:

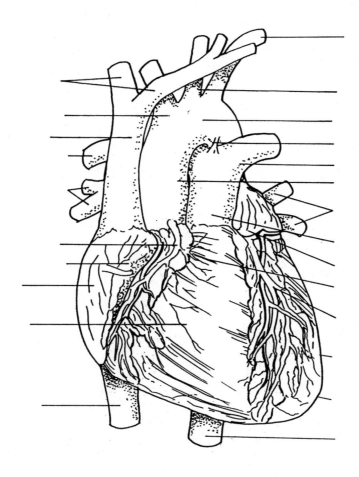

Figure 20-5

NOTES:

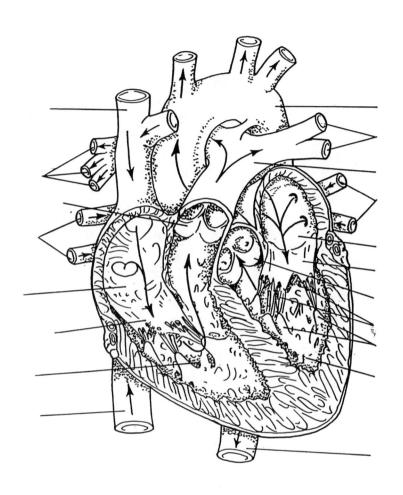

F i g u r e 20 - 6

NOTES:

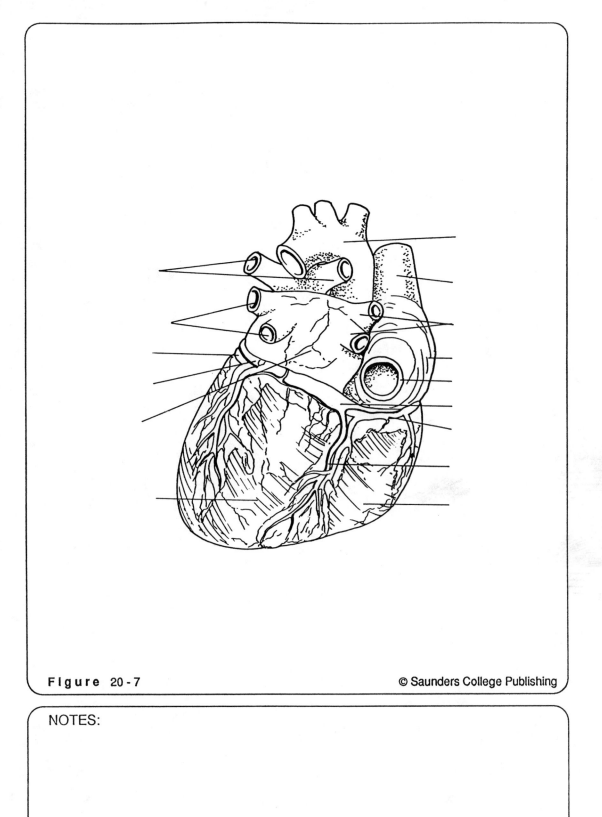

Figure 20-7

NOTES:

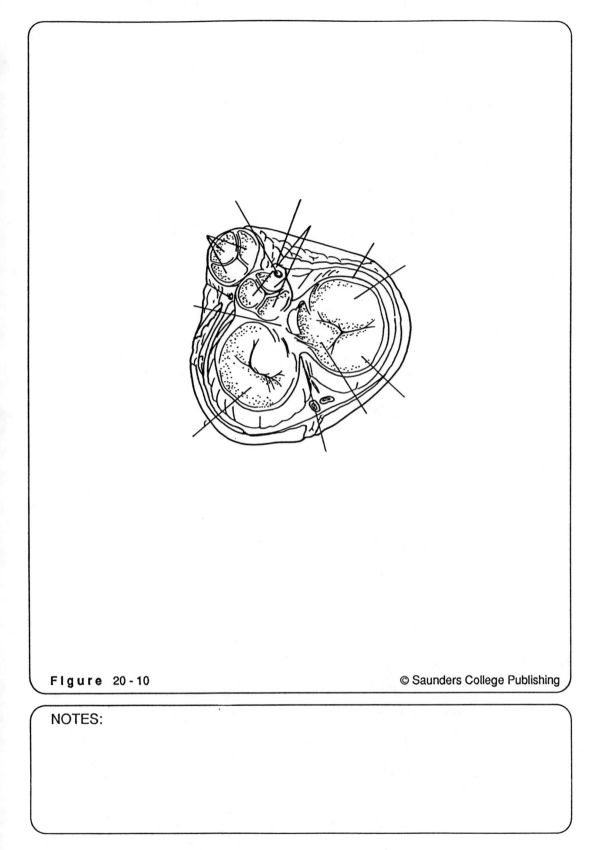

F i g u r e 20 - 10

NOTES:

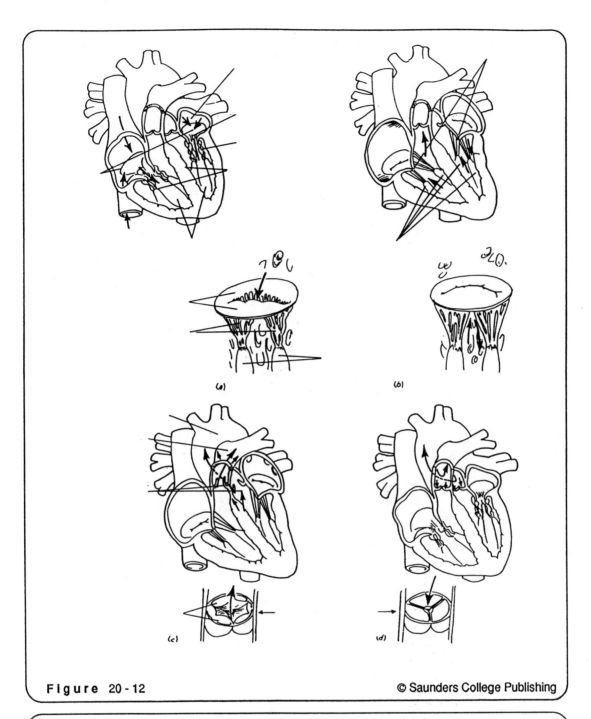

Figure 20-12

NOTES:

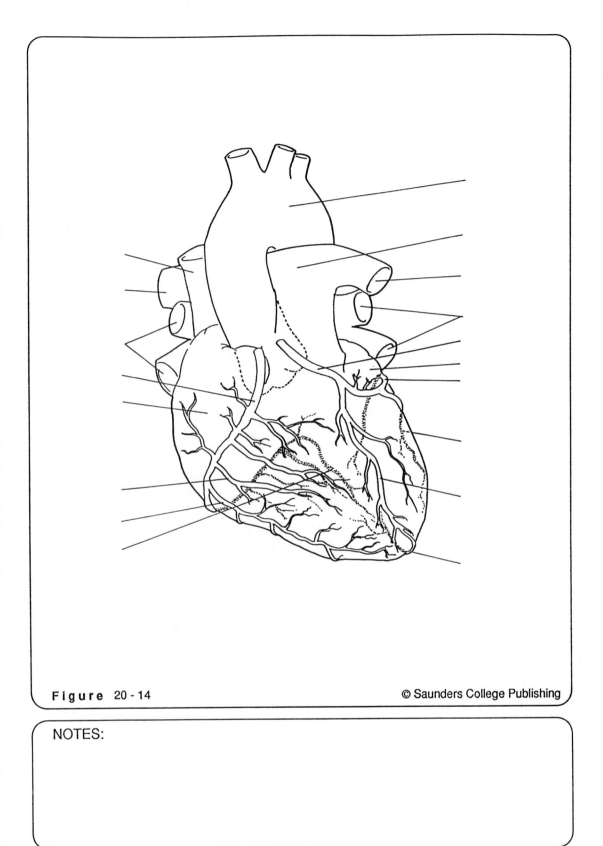

Figure 20 - 14

NOTES:

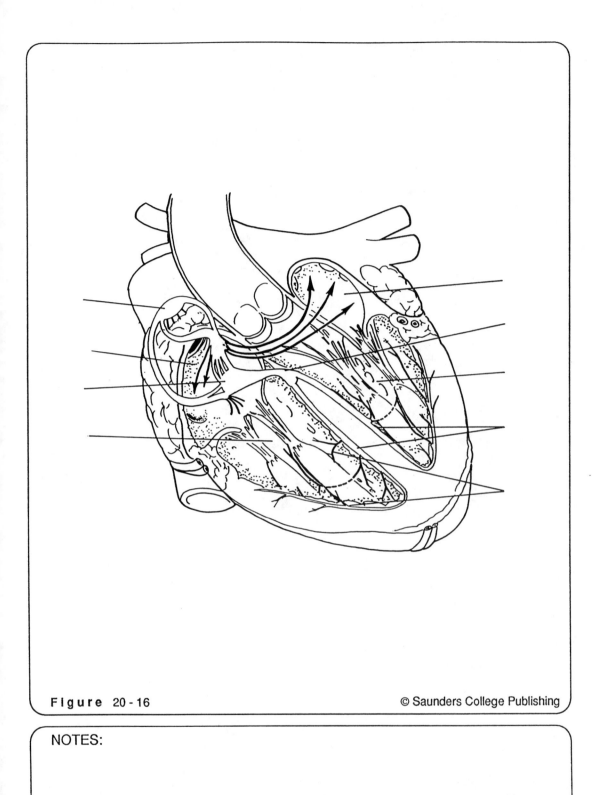

F i g u r e 20 - 16

NOTES:

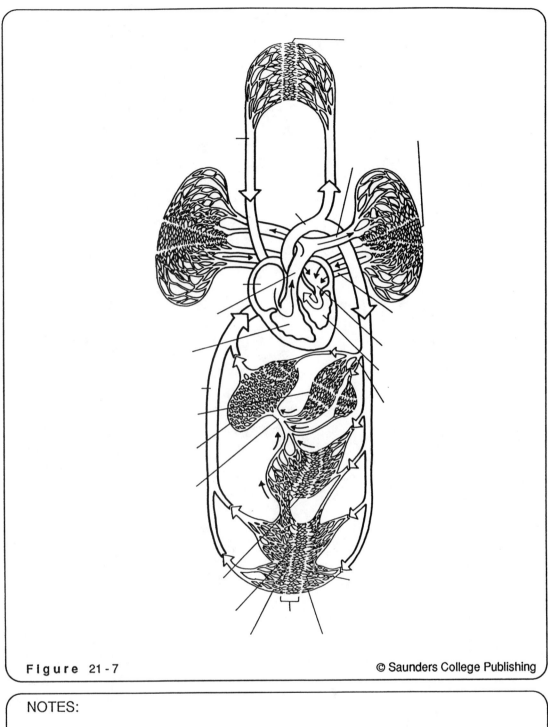

Figure 21-7

NOTES:

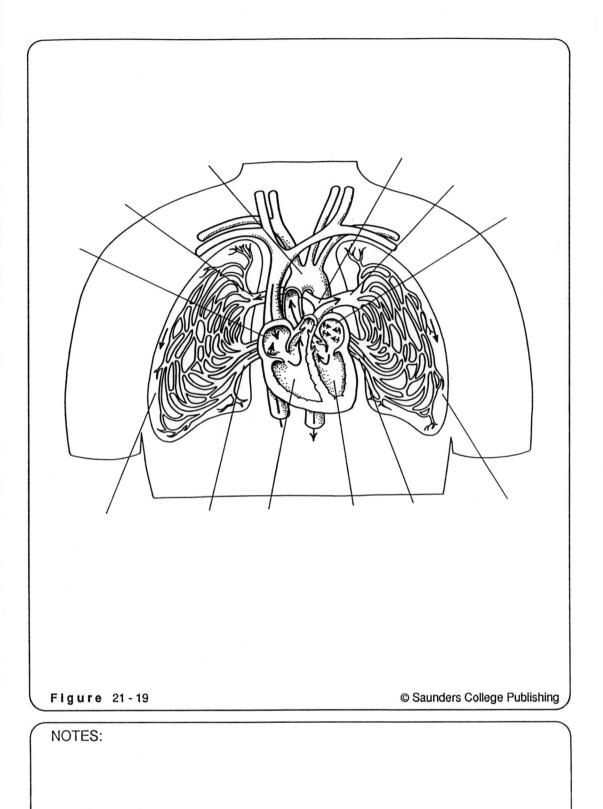

Figure 21 - 19

© Saunders College Publishing

NOTES:

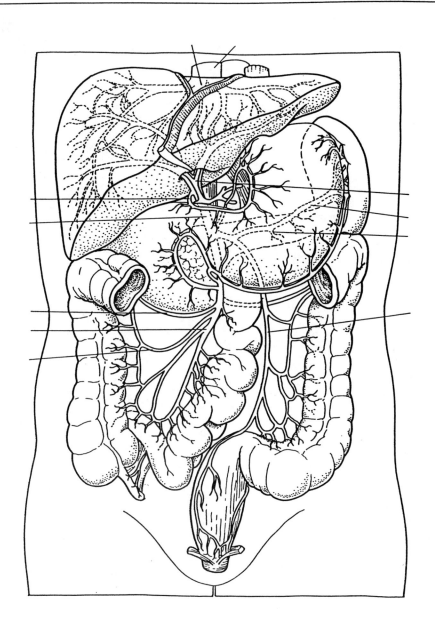

Figure 21-36

NOTES:

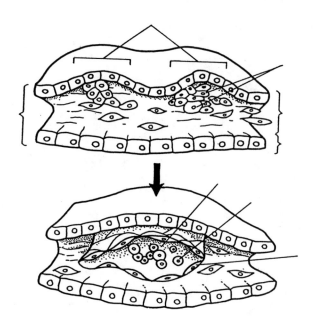

Figure 21-42

NOTES:

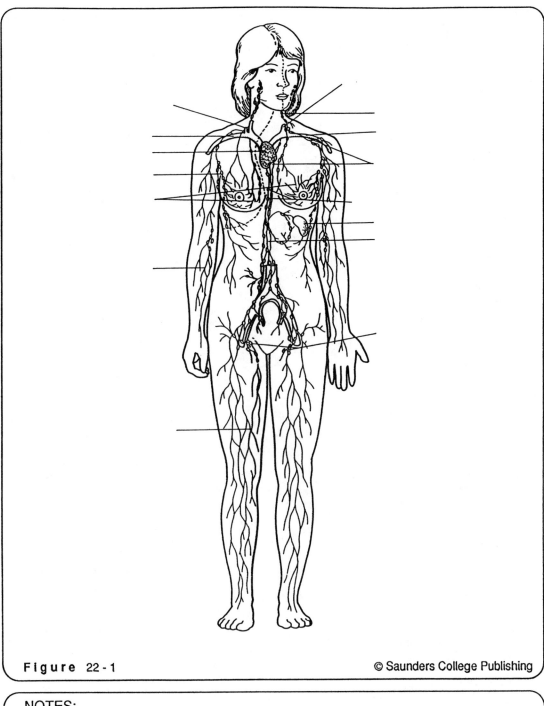

Figure 22-1

NOTES:

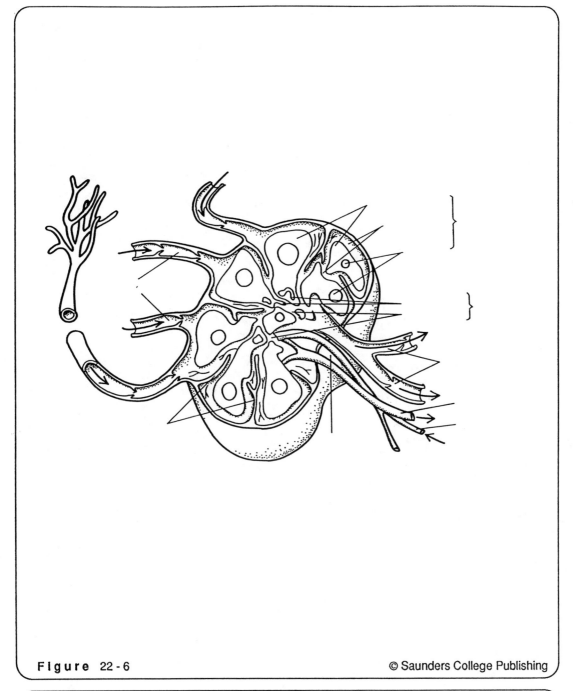

Figure 22 - 6

NOTES:

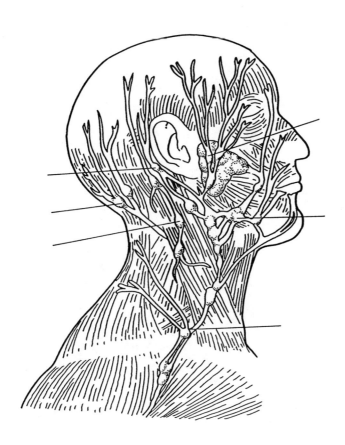

Figure 22-7

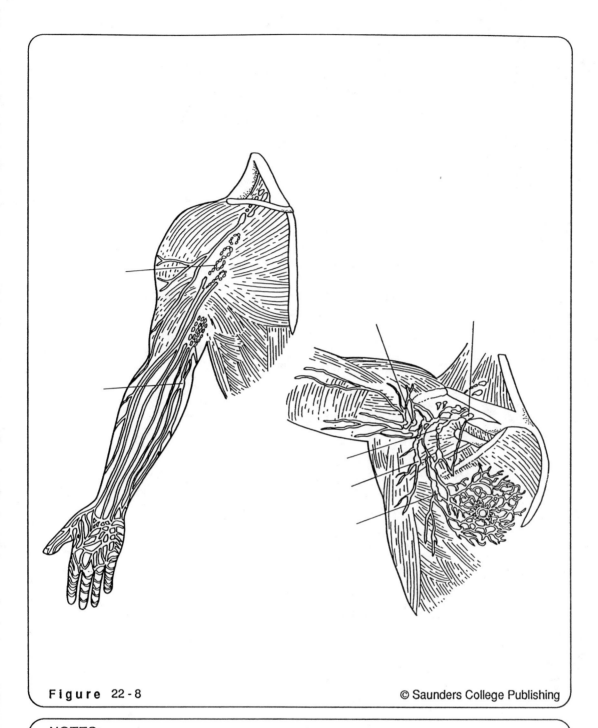

Figure 22-8

© Saunders College Publishing

NOTES:

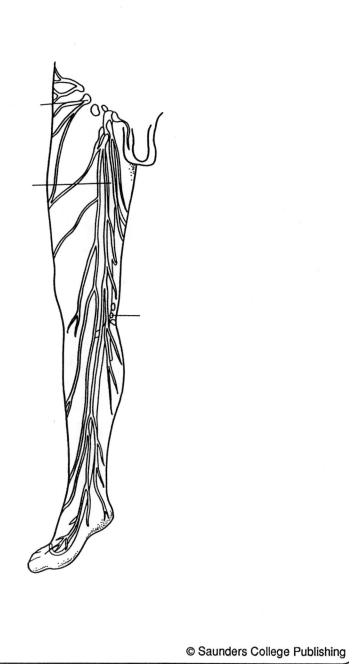

Figure 22-9

NOTES:

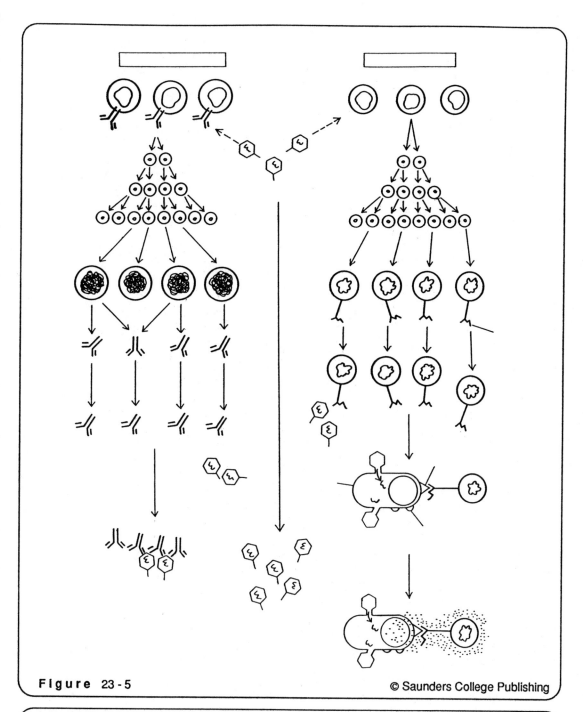

Figure 23-5

NOTES:

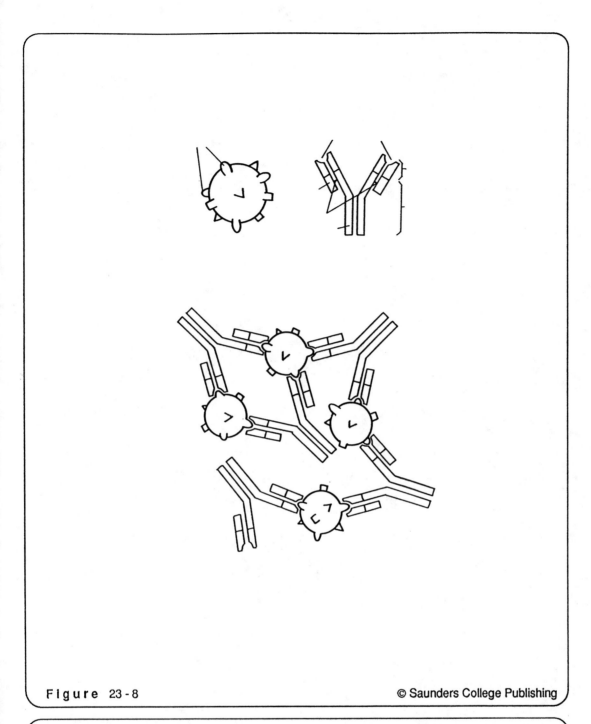

Figure 23-8   © Saunders College Publishing

NOTES:

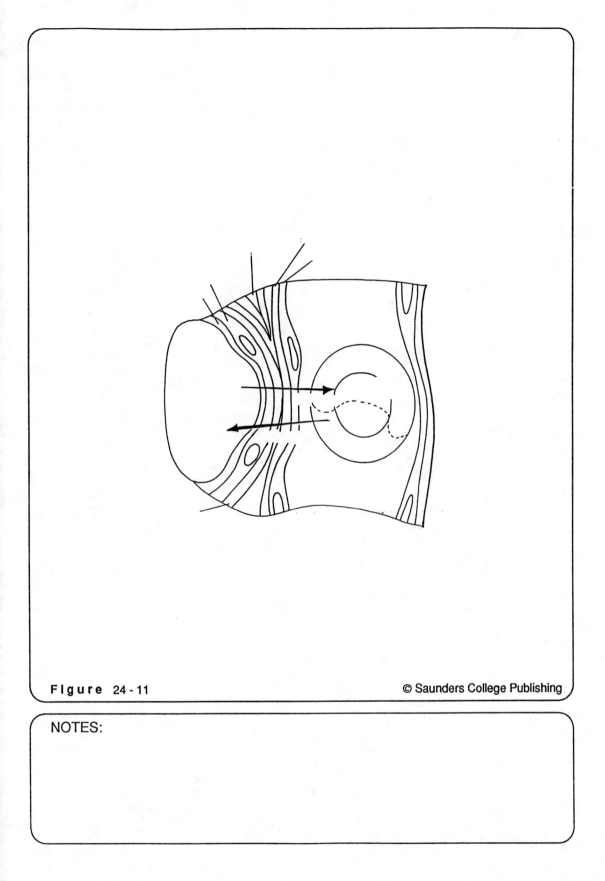

Figure 24 - 11

NOTES:

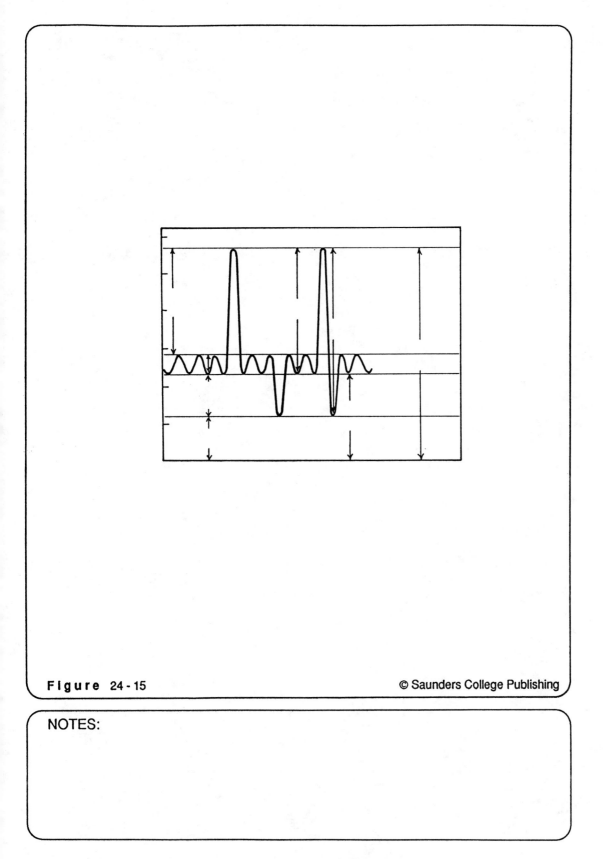

Figure 24-15

NOTES:

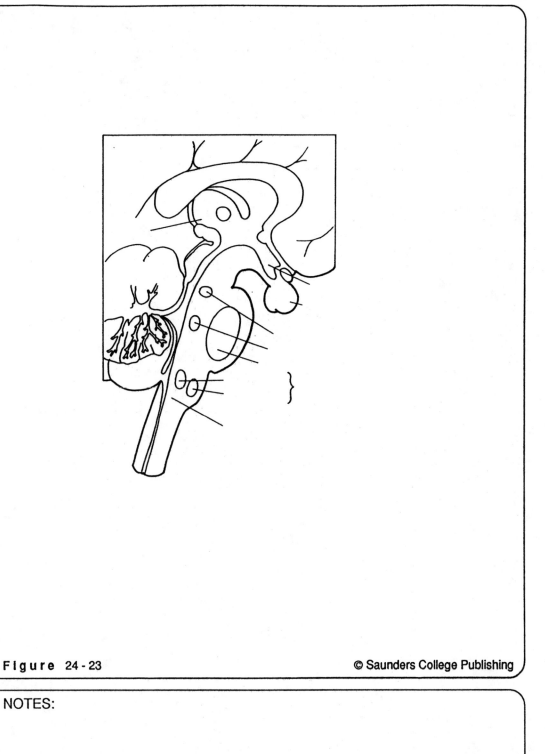

Figure 24 - 23

NOTES:

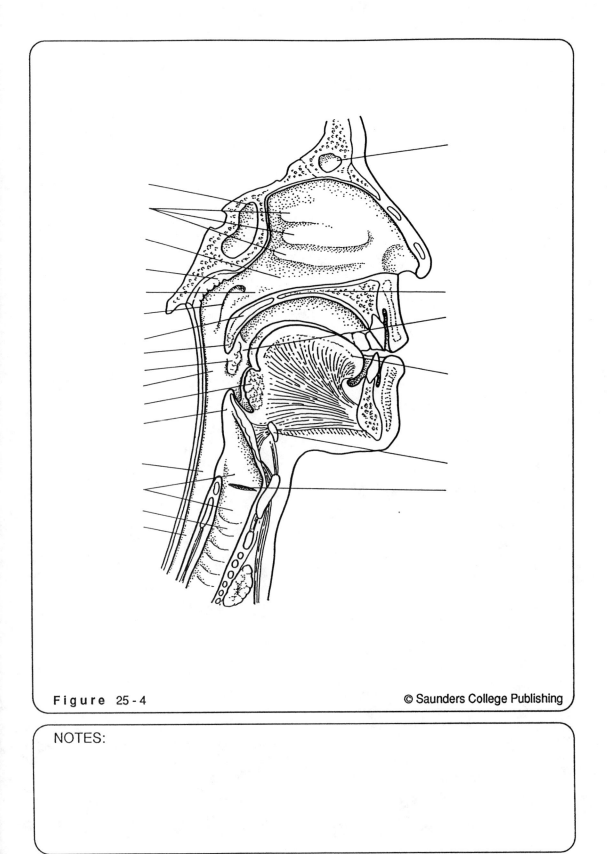

Figure 25 - 4

NOTES:

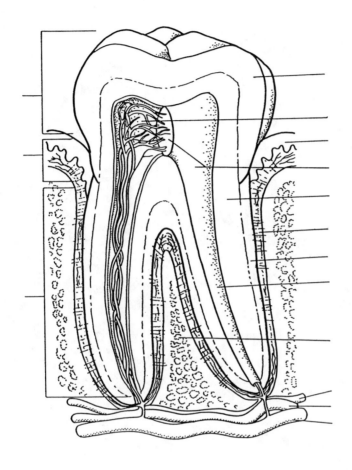

Figure 25-6

NOTES:

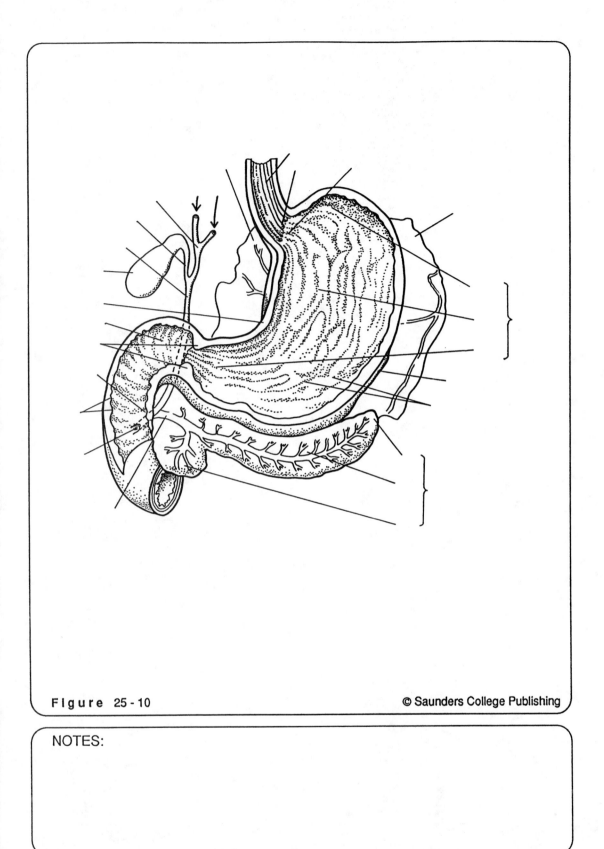

Figure 25-10

© Saunders College Publishing

NOTES:

105

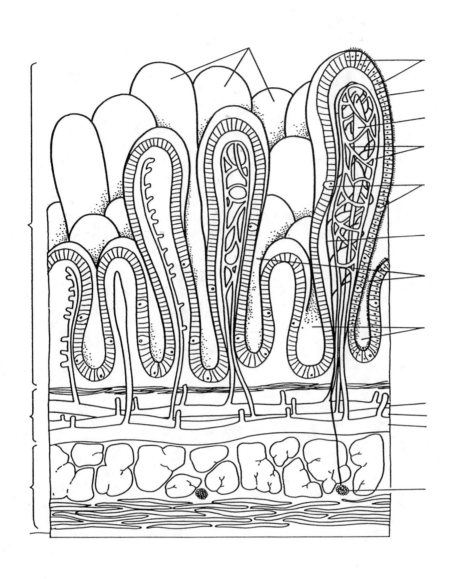

Figure 25-21

© Saunders College Publishing

NOTES:

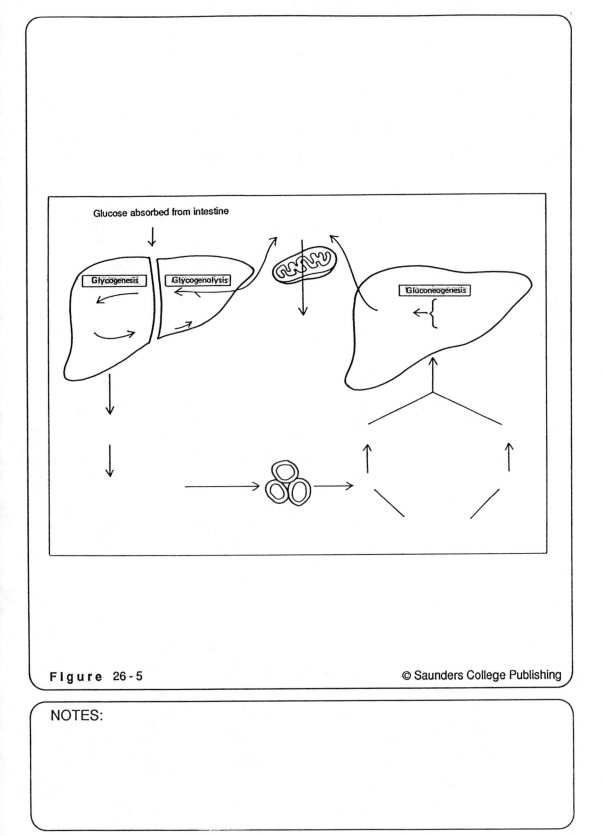

Figure 26-5

NOTES:

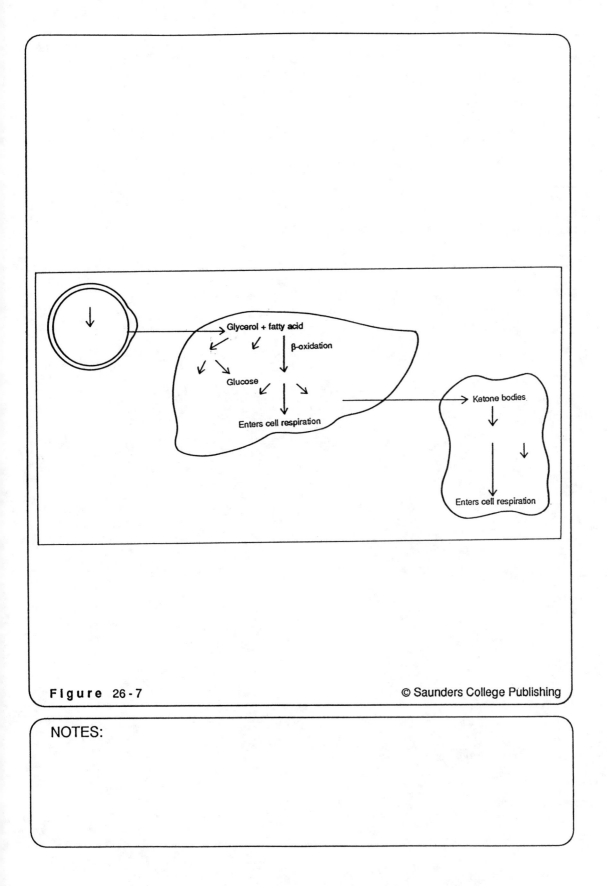

Glycerol + fatty acid

β-oxidation

Glucose

Enters cell respiration

Ketone bodies

Enters cell respiration

F i g u r e 26 - 7

© Saunders College Publishing

NOTES:

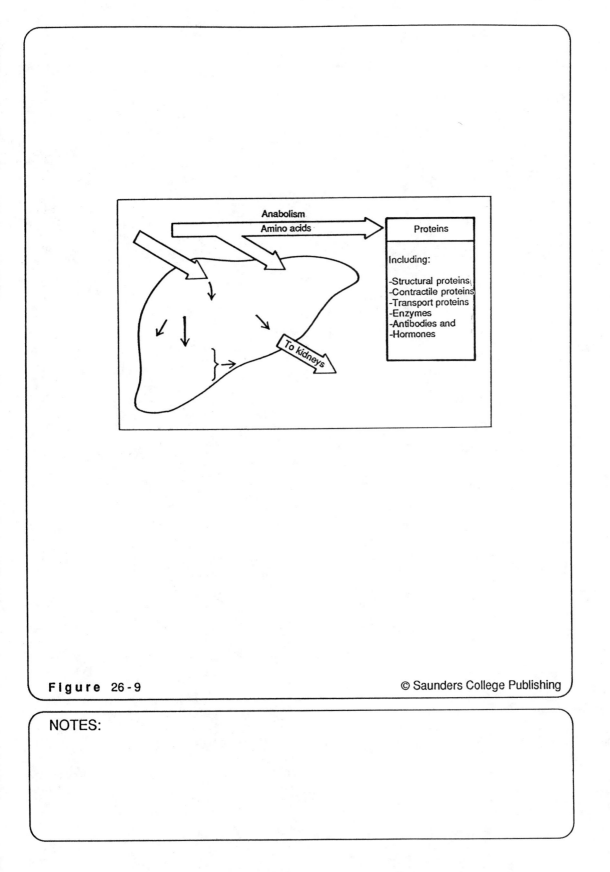

Figure 26-9

© Saunders College Publishing

NOTES:

109

Figure 28-8

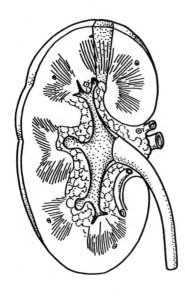

Figure 27-3

NOTES:

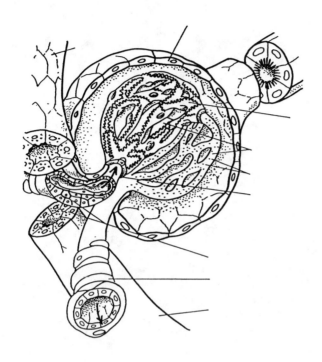

Figure 27 - 4a

NOTES:

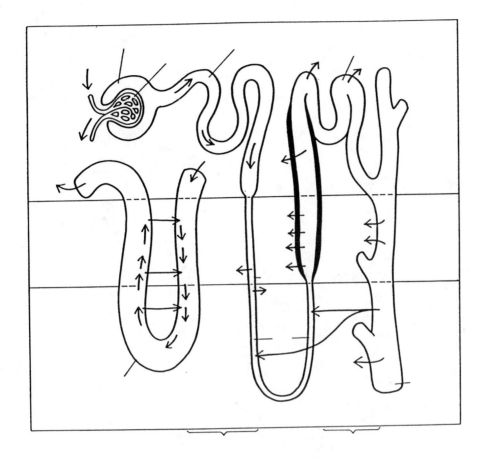

Figure 27-16

NOTES:

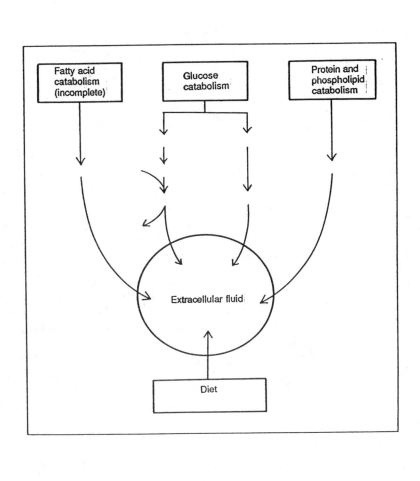

Figure 28 - 6

NOTES:

113

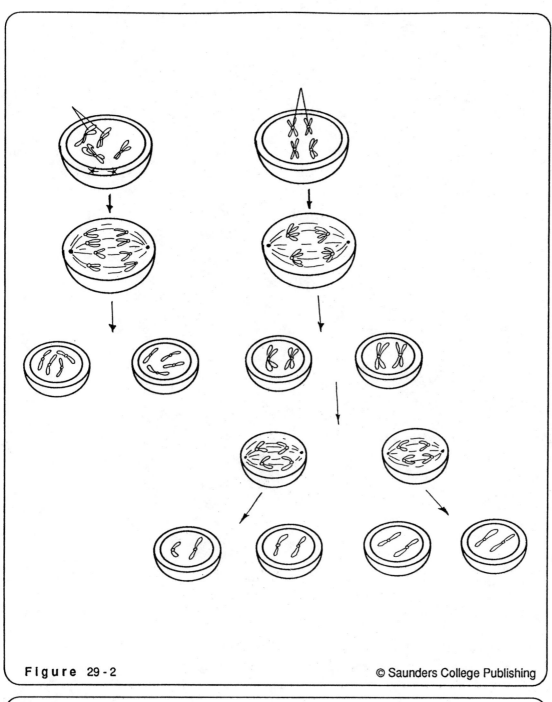

Figure 29-2

NOTES:

Figure 21-2

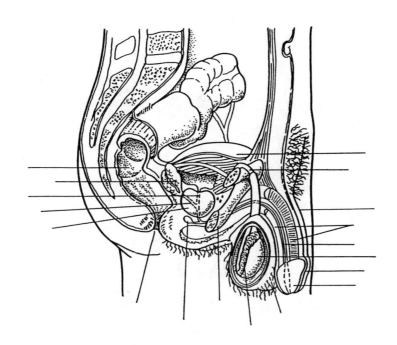

Figure 29-4

NOTES:

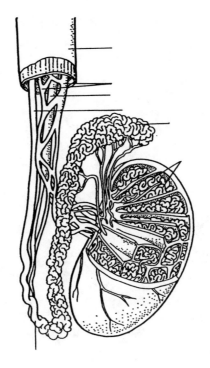

F i g u r e 29 - 5b

NOTES:

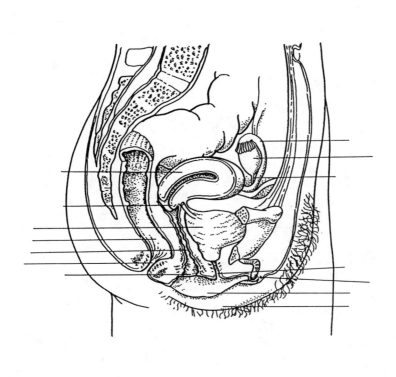

F i g u r e 29 - 14

NOTES:

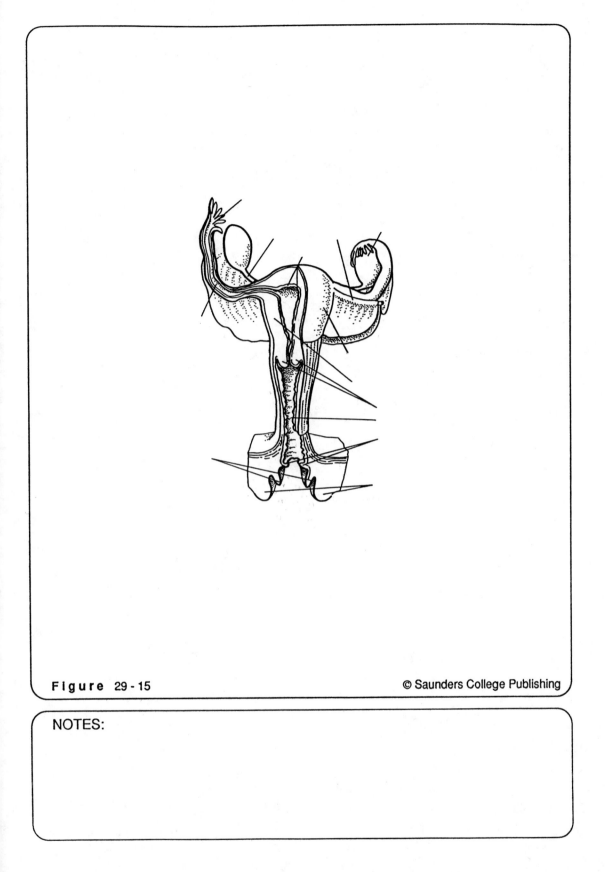

F i g u r e 29 - 15

NOTES:

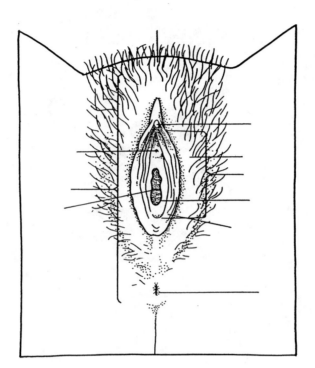

Figure 29-20

NOTES:

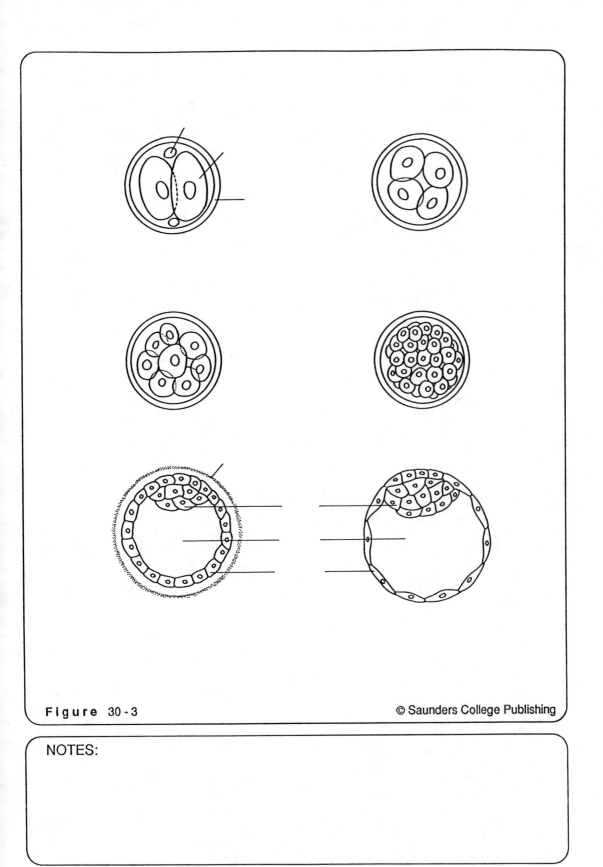

Figure 30-3

NOTES:

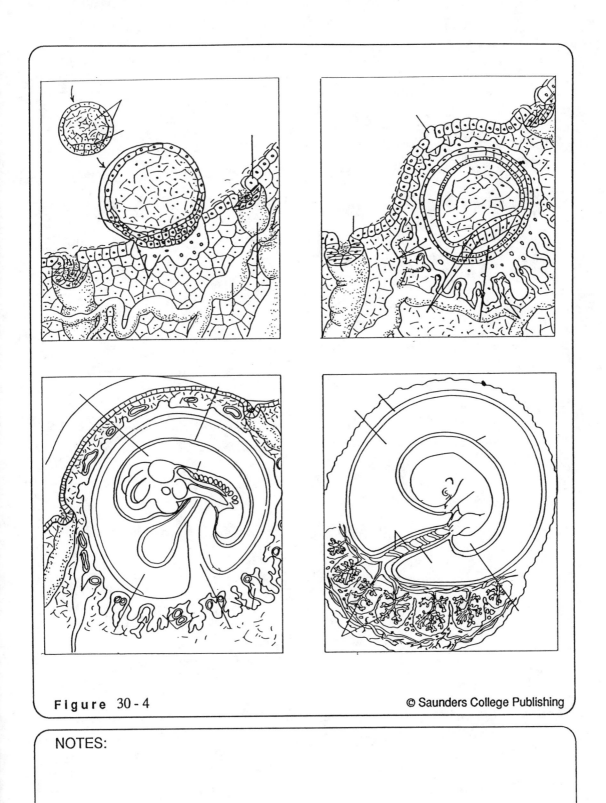

Figure 30-4

NOTES:

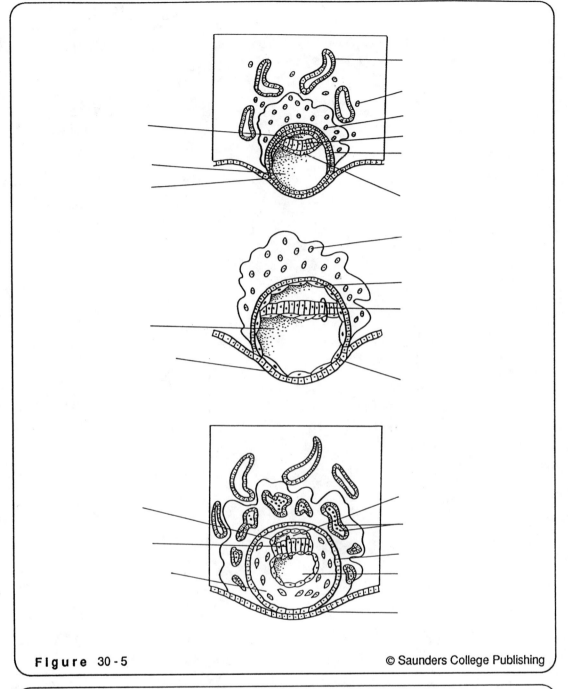

Figure 30-5

NOTES:

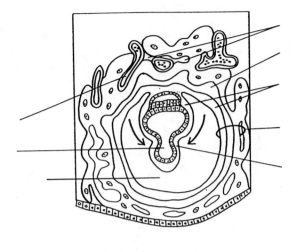

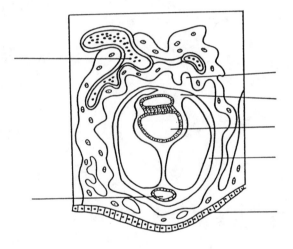

© Saunders College Publishing

Figure 30 - 7

NOTES:

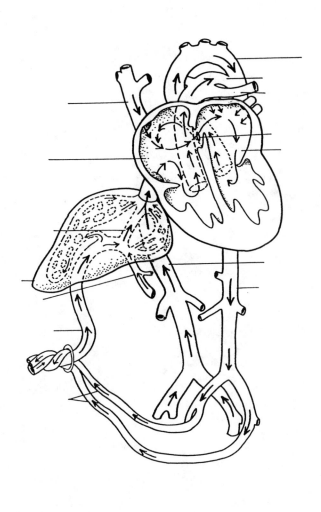

Figure 30 - 18

NOTES:

Figure 3-2

O Chromatin
O Plasma membrane
O Rough endoplasmic reticulum
O Mitochondria
O Zymogen granules
O Ribosomes

O Desmosome
O Golgi complex
O Smooth endoplasmic reticulum
O Nucleus
O Nucleolus

Window on the Animal Cell

O Plasma membrane
O Cytoplasm
O Microfilaments
O Rough endoplasmic reticulum
O Smooth endoplasmic reticulum
O Pores of the nuclear membrane
O Ribosomes
O Nucleoplasm

O Nucleolus
O Centrioles
O Lysosome
O Golgi complex
O Vesicles of the Golgi complex
O Microbodies
O Pinocytotic vesicle
O Microvilli

Window on the Human Body,
View #4

O Larynx
O Thyroid gland
O Trachea
O Deltoid Muscle
O Left Lung
O Right Lung
O Aortic arch
O Diaphragm
O Stomach
O Transverse colon
O Coils of the small intestine
O Inguinal ligament
O Tensor fasciae latae muscle
O Urinary bladder

O Sartorious muscle
O Scrotum
O Femoral artery
O Femoral vein
O Appendix
O Caecum
O Duodenum
O Hepatic flexure
O Gall bladder
O Liver
O Right oblique fissure
O Horizontal fissure
O Heart in the pericardial sac
O Falciform ligament

Figure 3-25

O Chromatin
O Centrioles
O Mitotic spindle
O Remnants of nuclear membrane
O Centromere
O Interphase
O Late interphase

O Cytokinesis
O Telophase
O Anaphase
O Early metaphase
O Early prophase
O Late prophase

Figure 4-2

O Unicellular gland
O Cilia
O Basement membrane
O Simple tubular
O Simple tubular with coiled tube
O Simple branched tubular

O Simple acinar
O Simple branched acinar
O Compound tubular
O Compound acinar
O Compound tubuloacinar

Figure 4-9

O Spongy bone
O Yellow marrow
O Compact bone
O Haversian canal
O Marrow cavity

O An osteon
O Lacuna osteocyte
O Cytoplasmic process
O Matrix

Figure 5-2

○ Hair shaft
○ Stratum corneum
○ Stratum granulosum
○ Stratum spinosum
○ Stratum basale
○ Arrector pili muscle
○ Sebaceous gland
○ Collagen fibers
○ Papilla of connective tissue
○ Hair follicle

○ Adipose tissue
○ Vein
○ Artery
○ Sweat gland
○ Dermal papilla
○ Sense organ
○ Nerve fiber
○ Epidermis
○ Dermis
○ Subcutaneous layer

Figure 6-1

○ Epiphysis
○ Hyaline articular cartilage
○ Compact bone
○ Epiphyseal lines
○ Cancellous bone trabeculae
○ Metaphysis

○ Diaphysis
○ Compact bone
○ Marrow cavity (medullary cavity)
○ Periosteum: osteogenic layer

Figure 6-5

○ Mesenchyme
○ Fibroblasts outside condensing area
○ Mesenchymal cells in random collagen matrix
○ Random collagen in dense matrix

○ Trabeculae
○ Fresh osteoid
○ Osteocyte in lacunae
○ Woven bone

Figure 6-10

○ Broken blood vessels
○ Hematoma
○ Broken periosteum
○ Marrow

Figure 6-15

○ Lacunae containing osteocytes
○ Cancellous bone
○ Endosteum
○ Blood vessels
○ Compact bone
○ Osteocyte processes in the canaliculi

○ Osteogenic layer in the periosteum
○ Endosteal lining of the Haversian canal
○ Fibrous layer of the periosteum

Figure 6-16

O Surface of the shaft
O Periosteum over the ridges
   forms bone to make the
   ridges higher
O Ridges meet and fuse
O Periosteum of the groove is now
   endosteum of the tunnel

O Layers of bone deposited inside
   the tunnel make a
   haversian system
O Haversian vessel, originally
   periosteal

Figure 6-19

O Bursa
O Muscle fibers
O Tendon
O Periosteum

O Sharpey's fibers
O Bone
O Muscle
O Bone

130

Figure 7-1

O Frontal
O Sternum
O Scapula
O Costal cartilages
O Xiphoid process
O "Floating rib"
O Lumbar vertebrae
O Ilium
O Sacrum
O Coccyx
O Pubis
O Ischium
O Femur
O Patella
O Tibia
O Fibula
O Talus
O Metatarsals

O Phalanges
O Parietal
O Nasal
O Temporal
O Orbit
O Maxilla
O Mandible
O Cervical vertebrae
O Clavicle
O "True ribs"
O Humerus
O "False ribs"
O Radius
O Ulna
O Carpals
O Metacarpals
O Phalanges
O Pubic symphysis

Figure 7-2

O Parietal
O Occipital
O Temporal
O Maxilla
O Zygomatic arch
O Mastoid process
O Mandible
O Clavicle
O Acromion
O Spine of scapula
O Scapula
O Humerus
O Radius
O Ulna
O Carpals
O Metacarpals
O Phalanges

O Cervical vertebrae
O Thoracic vertebrae
O Lumbar vertebrae
O Iliac crest
O Sacrum
O Head of femur
O Coccyx
O Ischium
O Femur
O Tibia
O Fibula
O Lateral malleolus
O Metatarsals
O Phalanges
O Talus
O Calcaneus

Figure 7-3 (a)

O Parietal
O Supraorbital foramen
O Temporal
O Sphenoid
O Ethmoid
O Lacrimal
O Zygomatic (malar)
O Maxilla
O Mandible
O Squama

O Superciliary arch
O Glabella
O Superior orbital fissure
O Nasal
O Perpendicular plate of ethmoid
O Middle concha
O Infraorbital foramen
O Inferior concha
O Vomer
O Mental foramen

Figure 7-4 (a)

O Coronal suture
O Parietal
O Squamous suture
O Squamous portion of temporal
O Temporal
O Lambdoidal suture
O Occipital
O External auditory meatus
O Moveable joint with mandible
O Mastoid process of temporal
O Styloid process of temporal

O Frontal
O Sphenoid (greater wing)
O Orbit
O Nasal
O Ethmoid
O Lacrimal
O Zygomatic process of temporal
O Infraorbital foramen
O Zygomatic (malar)
O Maxilla

Figure 7-6 (a)

O Frontal
O Crista galli
O Optic canal
O Foramen rotundum
O Foramen ovale
O Internal auditory meatus
O Jugular foramen
O Hypoglossal canal
O Foramen magnum
O Cribriform plate of ethmoid

O Lesser wing of sphenoid
O Greater wing of sphenoid
O Sella turcica
O Temporal
O Foramen spinosum
O Foramen lacerum
O Petrous portion of temporal
O Parietal
O Occipital

132

Figure 7-8

O Incisive foramen
O Palatine process of maxilla
O Horizontal plate of palatine
O Vomer
O Foramen ovale
O Foramen lacerum
O Carotid canal
O Jugular fossa
O Occipital condyle
O Foramen magnum
O Occipital
O Parietal

O Inferior and superior nuchal
   lines
O Alveoli (tooth sockets)
O Zygomatic process of maxilla
O Zygomatic (malar)
O Lateral pterygoid plate
O Medial pterygoid plate
O Greater wing
O Styloid process
O Mastoid process
O Stylomastoid foramen
O Median nuchal crest

Figure 7-9

O Coronal suture
O Bregma
O Lambdoidal suture
O Lambda
O Sagittal suture
O Bregma

Figure 7-12

O Spine
O Vertebral foramen
O Pedicle
O Centrum (body)
O Lamina
O Superior articular process
O Transverse foramen
O Transverse process
O Superior transverse process
O Centrum (body)
O Facet for tubercle of rib
O Superior facet for head of rib

O Inferior articular process
O Superior articular facet
O Cervical vertebrae
O Thoracic vertebrae
O Lumbar vertebrae
O Sacral vertebrae
O Coccygeal vertebrae
O Cervical curve
O Thoracic curve
O Lumbar curve
O Sacral curve

133

Figure 7-13

O Superior articular facets
O Tuberosity
O Median crest
O Hiatus
O Sacral canal
O Lateral crest
O Dorsal sacral foramina
O Coccyx
O Superior articularing process
O Transverse lines
O Articular surface

O Promontory
O Pelvic sacral foramina
O Superior articular facet
O Costal facets for articulation
    with tubercles of ribs
O Spine
O Inferior articular facet
O Superior facet
O Inferior facet
O Centrum
O Intervertebral foramen
O Position of vertebral canal

Figure 7-15 (a)

O Superior articular facet
O Transverse Process
O Anterior arch
O Anterior tubercule
O Odontoid process of the axis

O Transverse foramen
O Vertebral foramen
O Posterior arch
O Posterior tubercule

Figure 7-15 (b)

O Transverse foramen
O Lamina
O Inferior articular facet
O Spine

O Facet for atlas
O Odontoid process
O Superior articular facet
O Centrum

Figure 7-15 (c)

O Odontoid process
O Atlas
O Transverse foramen
O Axis

134

Figure 7-18

○ First rib
○ Costal cartilages
○ Seventh rib
○ Eleventh and Twelfth
    ribs (floating ribs)

○ First thoracic vertebra
○ Manubrium
○ Body
○ Xiphoid

Figure 7-18 (b)

○ Angle
○ Neck
○ Vertebrae
○ Shaft
○ Rib
○ Costal cartilage

○ Xiphoid process
○ Body
○ Sternum
○ Rib articulations
○ Clavicular notch
○ Manubrium

Figure 8-2 (a)

O Superior angle
O Vertebral border
O Inferior angle
O Superior border
O Spine

O Coracoid process
O Acromion process
O Glenoid fossa
O Axillary border
O Dorsal surface

Figure 8-3 (a)

O Acromial end
O Sternal end

Figure 8-3 (b)

O Articular surface for acromion
O Trapezoid line
O Conoid tubercle
O Articular surface for first costal
    cartilage
O Articular surface for sternum

Figure 8-4

O Greater tubercle
O Head
O Intertubercular groove
O Lesser tubercle
O Deltoid tuberosity
O Nutrient foramen
O Lateral epicondyle

O Coronoid fossa
O Medial epicondyle
O Capitulum
O Trochlea
O Anatomic neck
O Surgical neck
O Olecranon fossa

Figure 8-6

- Olecranon process
- Radial head
- Neck
- Radial tuberosity
- RADIUS
- Interosseus membrane
- Styloid process

- Carpal surfaces
- Coronoid process
- Ulnar tuberosity
- ULNA
- Interosseus borders
- Styloid process

Figure 8-9

- Iliac fossa
- Anterior superior iliac spine
- Anterior inferior iliac spine
- Pubic tubercle
- Pubic symphysis
- Superior ramus of pubis
- Inferior ramus of pubis
- Iliac crest
- Sacru

- Posterior inferior iliac spine
- Greater sciatic notch
- Ilium
- Ischial spine
- Lesser sciatic notch
- Acetabulum
- Ischial tuberosity
- Ishium

Figure 8-13

- Head
- Greater trochanter
- Neck
- Intertrochanteric crest
- Intertrochanteric line
- Lesser trochanter
- Spiral line
- Linea aspera

- Lateral epicondyle
- Lateral condyle
- Medial epicondyle
- Lateral supracondylar ridge
- Medial condyle
- Intercondylar fossa
- Lateral condyle

Figure 8-15

○ Intercondylar tubercles
○ Lateral condyle
○ Fibular head
○ FIBULA
○ Lateral surface
○ Anterior border
○ Lateral malleolus
○ Medial condyle

○ Tibial tubercle
○ TIBIA
○ Interosseus border
○ Anterior tibial border (crest)
○ Medial surface
○ Anterior surface
○ Medial malleolus
○ Inferior articular surface

Figure 8-18

○ Distal phalanges
○ Middle phalanges
○ Metatarsals
○ Cuneiforms
○ Navicular
○ Talus
○ Calcaneus
○ Proximal phalanges
○ Cuboid
○ Tibia
○ Fibula
○ Lateral malleolus

○ Talus
○ Navicular
○ Calcaneus
○ Cuneiforms
○ Cuboid
○ Medial longitudinal arch
○ Metatarsals
○ Lateral longitudinal arch
○ Transverse arch
○ Proximal phalanges
○ Distal phalanges

Figure 9-1

O Suprapatellar bursa
O Tendon of quadriceps femoris
O PATELLA
O Subcutaneous prepatellar bursa
O Infrapatellar fat pad

O Deep infrapatellar bursa
O Fibrous capsule
O Posterior cruciate ligament
O Anterior cruciate ligament

Figure 9-2

O Subacromial bursa
O Deltoid
O Glenoid cavity
O Subscapularis
O Pectoralis major
O Musclocutaneous nerve
O Coracobrachialis and biceps-
    short head
O Axillary artery
O Ulnar nerve
O Axillary vein

O Tendon of latissimus dorsi
O Acromion
O Articular capsule
O Deltoid
O Glenoidal labrum
O Teres minor
O Axillary nerve
O Radial nerve
O Triceps, long head
O Teres major

Figure 9-3

O Flexor pollicis longus
O Muscles of thenar eminence
O Flexor carpi radialis
O Trapezium
O First metacarpal bone
O Extensor pollicis brevis
O Extensor pollicis longus
O Radial artery
O Extensor carpi radialis longus
O Trapezoid bone
O Extensor carpi radialis brevis
O Flexor digitorum profundus
O Flexor retinaculum

O Ulnar artery
O Ulnar nerve
O Flexor digitorum superficialis
O Muscles of the hypothenar
    eminence
O Hamate bone
O Extensor carpi ulnaris
O Extensor digiti minimi
O Capitate bone
O Tendons of the extensor
    digitorum and extensor
    indicis

Figure 9-6

O Enamel
O Dentinal tubules
O Pulp chamber
O Odontoblasts
O Alveolar mucosa
O Cement
O Pulp canal
O Predentine
O Apical foramen
O Dentine

O Gingival sulcus
O Gingival fibers
O Gingiva
O Periodontal ligament
O Alveolar bone
O Penetrating Sharpley's fibers
O Basal bone
O Nutrient vessel

Figure 9-8

O Hyaline cartilage
O Superior pubic ligament
O Interpubic disc of fibrocartilage
O Body of the pubis
O Arcuate ligament

Figure 9-9

O Nucleus pulposus
O Annulus fibrosus
O Anterior longitudinal ligament
O Intervertebral disc
O Ligamentum flavum

O Lamina
O Interspinous ligament
O Supraspinous ligament
O Pedicle

140

Figure 9-13

○ Ligament of Wrisberg
○ Medial condyle
○ Medial meniscus
○ Posterior cruciate ligament
○ Tibial collateral ligament
○ TIBIA

○ FEMUR
○ Anterior cruciate ligament
○ Lateral condyle
○ Lateral meniscus
○ Fibular collateral ligament
○ FIBULA

Figure 9-16

○ Patellar surface
○ Lateral condyle
○ Lateral meniscus
○ Fibular collateral ligament
○ Fibula
○ Medial condyle
○ Posterior cruciate ligament

○ Anterior cruciate ligament
○ Coronary ligament
○ Medial meniscus
○ Transverse ligament
○ Tibia collateral
   ligament
○ Tibia

Chapter 9-p.285 (b)

- Posterior tibiotalar
- Tibiocalcanean
- Tibionavicular
- Sustentaculum tali
- Medial maleolus

- Talonavicular ligament
- Long plantar ligament
- Tuberosity of the navicular bone
- First metatarsal bone

Chapter 9-p.285 (c)

- Anterior tibiofibular ligament
- Anterior talofibular ligament
- Talonavicular ligament
- Posterior tibiofibular ligament
- Posterior talofibular ligament

- Calcaneofibular ligament
- Bifurcated ligament
- Long plantar ligament
- Cervical ligament

Chapter 9-p.287 (f)

- Anterior inferior iliac spine
- Iliofemoral ligament
- Greater trochanter
- Intertrochanteric line

- Iliopectineal eminence
- Pubofemoral ligament
- Obturator foramen
- Lesser trochanter

Chapter 9-p.289

- Gluteus medius
- Gluteus minimus
- Piriformis
- Gluteus maximus
- Sciatic nerve
- Posterior femoral muscles
- Tensor fasciae latae
- Sartorius
- Acetabular labrum

- Articular capsule
- Transverse acetabular ligament
- Medial circumflex femoral artery
- Obturator nerve posterior division
- Adductor brevis
- Gracilis
- Adductor magnus, ischial fibers

Figure 10-1

O Muscle (biceps femoris)
O Epimysium
O Perimysium
O Muscle fascicle
O Endomysium

O Muscle cell (muscle fiber)
O Sarcolemma
O Myofibril
O Myofilaments

Figure 10-3

O Sarcoplasmic reticulum
O Pore
O Sarcolemma
O Mitochondria
O Cell nucleus
O H-band
O M-band

O I-band
O Myofilaments
O Z-line
O Sarcomere
O A-band
O Terminal cistern
O T-tubule

Figure 10-6

O Myofibril
O Filaments
O Cross bridges
O Actin (thin filament)
O Myosin (thick filament)
O Sarcomere
O A band

O I band
O H zone
O Z line
O Cross bridges
O Thin filaments (actin)
O Thick filaments (myosin)

143

Figure 10-8

O Motor neuron
O End plate
O Myoneural cleft
O Transmitter substance
O Depolarized area of sarcolemma

Figure 10-9

O Vesicle of sarcoplasmic
      reticulum
O Calcium

Figure 10-10

O Stimuli
O Simple twitch

Figure 10-14

O Myosin head
O Actin filaments
O calcium activated site on the
      actin

Figure 10-17 (a)

O Nerve cell
O Smooth muscle cells

Figure 11-4

- Flexor digitorum superficialis
- Flexor pollicis longus
- Extensor carpi ulnaris
- Latissimus dorsi
- Rectus abdominis
- Linea alba
- External oblique
- Gluteus medius
- Iliopsoas
- Adductor longus
- Gracilis
- Adductor magnus
- Sartorius
- Vastus lateralis
- Quadriceps femoris
- Vastus medialis
- Patella
- Patellar ligament
- Tibialis anterior
- Peroneus longus
- Soleus
- Platysma
- Orbicularis oculi

- Zygomaticus
- Orbicularis oris
- Sternocleidomastoid
- Levator scapulae and scalenes
- Trapezius
- Clavicle
- Deltoid
- Pectoralis major
- Biceps brachii
- Serratus anterior
- Triceps brachii
- Brachialis
- Pronator teres
- Brachioradialis
- Flexor carpi radialis
- Flexor carpi ulnaris
- Tensor fasciae latae
- Gastrocnemius
- Peroneus longus
- Extensor digitorum longus
- Tibialis anterior
- Tibia
- Flexor digitorum longus

Figure 11-5

- Opponens pollicis
- Flexor pollicis longus
- Flexor carpi ulnaris
- Flexor digitorum superficialis
- Brachioradialis
- Pectoralis major
- Internal intercostal
- Sartorius
- Iliopsoas
- Gluteus medius
- Adductor longus
- Adductor brevis
- Adductor magnus
- Gracilis
- Vastus lateralis
- Vastus intermedius
- Quadriceps femoris
- Tendon of rectus femoris
- Vastus medialis
- Gastrocnemius
- Tibialis anterior
- Temporalis
- Corrugator supercilii
- Orbicularis oculi
- Zygomaticus

- Masseter
- Orbicularis oris
- Scalenes
- Trapezius
- Sternocleidomastoid
- Long head of the biceps
- Short head of the biceps
- Pectoralis minor
- Coracobrachialis
- Serratus anterior
- Medial head of the triceps
- Brachialis
- Rectus abdominis
- Transversus abdominis
- Internal oblique
- Brachioradialis
- External oblique
- Tensor fasciae latae
- Fascia lata
- Femur
- Patella
- Tibia
- Fibula
- Peroneus brevis

Figure 11-6

- O Orbicularis oculi
- O Sternocleidomastoid
- O Zygomaticus
- O Masseter
- O Buccinator
- O Splenius capitis
- O Trapezius
- O Deltoid
- O Triceps brachii
- O Brachioradialis
- O Palmaris longus
- O Flexor carpi radialis
- O Flexor pollicis longus
- O Flexor digitorum superficialis
- O Biceps brachii
- O Extensor carpi ulnaris
- O Extensor digitorum
- O Extensor carpi radialis brevis

- O Brachialis
- O Infraspinatus
- O Teres minor
- O Teres major
- O Rhomboideus major
- O Latissimus dorsi
- O External oblique
- O Gluteus maximus
- O Adductor magnus
- O Gracilis
- O Semitendinosus
- O Biceps femoris
- O Semimembranosus
- O Gastrocnemius
- O Soleus
- O Achilles tendon
- O Peroneus brevis
- O Calcaneus

Figure 11-7

- Semispinalis capitis
- Longissimus capitis
- Splenius capitis
- Trapezius
- Longissimus cervicis
- Iliocostalis cervicus
- Deltoid
- Supraspinatus
- Infraspinatus
- Teres minor
- Teres major
- Transversus abdominus
- Gluteus minimus
- Piriformis
- Adductor magnus
- Tibialis posterior
- Peroneus longus
- Peroneus brevis
- Levator scapulae
- Extensor pollicis longus
- Extensor carpi radialis brevis

- Extensor carpi radialis longus
- Brachioradialis
- Biceps brachii
- Triceps brachii
- Rhomboideus minor
- Rhomboideus major
- Latissimus dorsi
- Longissimus thoracis
- External intercostal
- Iliocostalis lumborum
- Internal oblique
- Erector spinae
- Gluteus medius
- Gluteus maximus
- Gracilis
- Semimembranosus
- Semitendinosus
- Biceps femoris
- Soleus
- Achilles tendon

Chapter 11-p.369

- ◯ Sternocleidomastoid
- ◯ Omohyoid (inferior belly)
- ◯ Pectoralis major
- ◯ Deltoid
- ◯ Triceps brachii
- ◯ Brachialis
- ◯ Biceps brachii
- ◯ Latissimus dorsi

- ◯ Serratus anterior
- ◯ External abdominal oblique
- ◯ Sternohyoid
- ◯ Trapezius
- ◯ Sternothyroid
- ◯ External intercostals
- ◯ Internal abdominal oblique
- ◯ Rectus abdominis

Chapter 11-p. 380

- ◯ Rectus femoris
- ◯ Gracilis
- ◯ Semitendinosus
- ◯ Vastus medialis
- ◯ Medial patellar retinaculum
- ◯ Patellar ligament
- ◯ Semimembranosus
- ◯ Gastrocnemius
- ◯ Soleus
- ◯ Tendo calcaneus
- ◯ Vastus intermedius

- ◯ Vastus medialis
- ◯ Sartorius
- ◯ Adductor magnus
- ◯ Gracilis
- ◯ Semimembranosus
- ◯ Semitendinosus
- ◯ Rectus femoris
- ◯ Femur
- ◯ Vastus lateralis
- ◯ Biceps femoris (short head)
- ◯ Beceps femoris (long head)

Chapter 11-p. 392 (a)

O Ischiocavernosus
O Bulbocavernosus
O Central tendon of the perineum
O Clitoris
O Urethral orifice
O Vagina

O Superficial transverse perineus
O Anus
O Gluteus maximus
O Pubococcygeus
O Iliococcygeus
O Levator ani

Chapter 11-p. 392 (b)

O Femoral head
O Obturator internus
O Vagina
O Bulbocavernosus

O Uterus
O Levator ani
O Urogenital diaphragm
O Ischiocavernosus

Chapter 11-p. 393 (c)

O Pubococcygeus
O Iliococcygeus
O Levator ani
O Ischiopubic ramus
O Bulbocavernosus

O Ischiocavernosus
O Superficial transverse perineus
O Central tendon of the perineum
O Gluteus maximus
O Anus

Chapter 11-p. 393 (d)

O Obturator internus
O Ischiocavernosus
O Levator ani
O Prostate
O Urogenital diaphragm
O Bulbocavernosus

Figure 12-3

O Cell body
O Nucleus
O Nissl substance
O Nodes of Ranvier
O Collateral branch
O Dendrites covered with
    dendritic spines

O Axon hillock
O Axon
O Myelin sheath
O Cellular sheath
O Axon sheath
O Synaptic knobs

Figure 13-1

O Skull
O Meninges
O Vertebra
O Spinal cord
O Meninges
O Epidural space
O Dura mater
O Subdural space
O Subarachnoid space
O Arachnoid

O Spinal cord
O Pia mater
O Dura mater
O Subarachnoid space
O Arachnoid
O Pia mater
O Spinal cord
O Bone of vertebra
O Epidural space
O Subdural space

Figure 13-7

O Thalamus
O Hypothalamus
O Diencephalon
O Pineal body
O Cerebral peduncles of midbrain
O Tentorium
O Cerebellum
O Fourth ventricle

O Pons
O Medulla
O Spinal cord
O Cerebrum
O Corpus callosum
O Fornix
O Optic chiasma
O Pituitary

Figure 13-9

O Cerebrum
O Infundibulum
O Tuber cinereum
O Mamillary body
O Hypothalamus
O Midbrain
O Pons
O Abducens nerve
O Roots of hypoglossal nerve
O Cerebellum
O Pyramids of medulla
O Longitudinal fissure

O Olfactory bulb
O Olfactory tract
O Optic nerve
O Optic chiasma
O Optic tract
O Oculomotor nerve
O Trochlear nerve
O Roots of trigeminal nerve
O Vestibulocochlear nerve
O Roots of facial nerve
O Glossopharyngeal nerve
O Roots of vagus nerve

Figure 13-21

O Cervical enlargement
O Lumbar enlargement
O Filum terminale of spinal cord
O Dura mater
O C1 (cervical nerve)
O C8
O T1 (thoracic nerve)
O T12

O L1 (lumbar nerve 1)
O Conus medullaris
O Cauda equina
O L5
O Dura mater
O S1 (sacral nerve 1)
O S5
O Coccygeal nerve

Figure 13-23

O Conus medullaris
O Filum terminale
O Spinal nerve roots
O Dura mater
O Nerve roots of first lumbar
      spinal nerve

Figure 13-24

O Central canal
O Lateral column
O Anterior white commissure
O Anterior median fissure
O Posterior (dorsal) fissure

O Posterior column
O Gray commissure
O Gray matter
O White matter
O Anterior column

Figure 14-1

○ Olfactory nerve (I)
○ Trochlear nerve (IV)
○ Oculomotor nerve (III)
○ Abducens nerve (VI)
○ Trigeminal nerve (V)
○ Optic nerve (II)
○ Facial Nerve (VII)
○ Vestibulocochlear nerve (VIII)
○ Spinal accessory nerve (XI)
○ V1

○ V2
○ V3
○ Pons
○ Vagus nerve (X)
○ Cerebellum
○ Glossopharyngeal nerve (IX)
○ Medulla
○ Hypoglossal nerve (XII)
○ Spinal cord

Figure 14-3

○ Dorsal gray horn
○ Dorsal white columns
○ Dorsal root filaments
○ Dorsal root
○ Dorsal root ganglion
○ Ventral root
○ Spinal nerve
○ Ventral root filaments

○ Spinal dura mater
○ Spinal arachnoid
○ Subarachnoid space
○ Spinal pia mater
○ Ventral white column
○ Ventral gray horn
○ Lateral white column

Figure 15-2

○ Medulla
○ Collateral nerve branch
○ Faciculus gracilis tract
○ First order neuron
○ Nucleus gracilis

○ Second order neuron
○ Thalamus
○ Third order neuron
○ Sensory area of the cerebral cortex
○ Nucleus in the thalamus

Figure 15-3

○ Intra-abdominal
○ Pharynx
○ Tongue
○ Teeth, gums and jaw
○ Lower lip
○ Lips
○ Upper lip
○ Genitals
○ Toes
○ Foot
○ Leg
○ Hip
○ Trunk
○ Neck
○ Head

○ Shoulder
○ Arm
○ Elbow
○ Forearm
○ Wrist
○ Hand
○ Little finger
○ Ring finger
○ Middle finger
○ Index finger
○ Thumb
○ Eye
○ Nose
○ Face

Figure 16-3

O Paravertebral ganglion
O Vagus nerve (parasympathetic)
O Heart
O Stomach
O Celiac ganglion

Figure 16-6

O Common carotid artery
O Carotid sinus
O Cardiac plexus
O Pulmonary plexus
O Esophageal plexus
O Celiac plexus
O Spleen
O Kidney

O Small intestine
O Colon
O Pancreas
O Stomach
O Liver
O Heart
O Lung

Figure 16-8

O Neck (C1-C4)
O Arms (C5-T3)
O Body Wall (T1-L3)
O Legs (L1-S3)
O Perineum (S4-S5)
O Superior cervical ganglion
O Pupil
O Salivary gland
O Larynx
O Trachea
O Lung
O Celiac ganglion
O Heart
O Gallbladder
O Adrenal gland

O Kidney
O Superior mesenteric ganglion
O Digestive system
O Inferior mesenteric ganglion
O Bladder
O Genitalia
O III
O VII
O IX
O X
O S1
O S2
O S3
O Sympathetic
O Parasympathetic

Figure 17-9(a)

O Sphenoid sinus
O Olfactory bulb
O Cribriform plate
O Frontal sinus
O Bundles of olfactory nerve fibers
O Inferior nasal concha
O Middle nasal concha

O Olfactory tract
O Second order neurons of the
    olfactory bulb
O Basal cells
O Olfactory cells
O Supporting cells
O Olfactory hairs

Figure 17-15(a)

O Pigmented epithelium
O Rod discs
O Rod cell
O Cone cell
O Horizontal cell
O Bipolar neuron

O Amacrine cell
O Optic nerve fibers
O Ganglionic neuron
O Vitreous body
O Retina
O Choroid layer and sclera

Figure 17-25

O Cochlear branch of the
    vestibulocochlear nerve
    (VIII)
O Vestbular branch of the
    vestibulocochlear nerve
    (VIII)
O Utricle
O Ampullae of the semicircular
    ducts
O Semicircular canals

O Semicircular ducts
O Saccule
O Osseous labyrinth of the cochlea
O Membraneous labyrinth
    of the cochlea
O Anterior canal
O Lateral canal
O Posterior canal

Figure 17-26 (a,b.c)

O Cochlear nerve
O Modiolus
O Spiral ganglion
O Organ of Corti
O Vestibular membrane
O Oval window
O Cochlear duct
O Walls of the membraneous
　　labyrinth

O Scala vestibuli
O Stria vascularis
O Scala media endolymph
O Tectorial membrane
O Basilar membrane
O Scala tympani
O Outer hair cells
O Supporting cells

Figure 18-2

- ○ Pituitary gland
- ○ Thymus gland
- ○ Pancreas
- ○ Ovary
- ○ Testis

- ○ Hypothalamus
- ○ Pineal gland
- ○ Thyroid gland
- ○ Parathyroid glands
- ○ Adrenal glands

Figure 20-1(a,b)

○ Heart
○ Diaphragm
○ Apex of the heart
○ Aorta
○ Superior vena cava

○ Pericardial cavity
○ Arch of the aorta
○ Pulmonary trunk
○ Parietal pericardium
○ Visceral pericardium

Figure 20-5

○ Brachiocephalic veins
○ Brachiocephalic artery
○ Superior vena cava
○ Right pulmonary artery
○ Right pulmonary veins
○ Right coronary artery
○ Auricle of the right atrium
○ Right atrium
○ Right ventricle
○ Inferior vena cava
○ Left subclavian artery
○ Left common carotid artery
○ ARCH OF THE AORTA

○ Ligamentum arteriosum
○ Left pulmonary artery
○ ASCENDING AORTA
○ Left pulmonary veins
○ Pulmonary artery
○ Auricle of the left atrium
○ Conus arteriosus
○ Descending branch of the left
    coronary artery
○ Left ventricle
○ Apex
○ DESCENDING AORTA

Figure 20-6

○ Superior vena cava
○ Aorta
○ Right pulmonary arteries
○ Pulmonary valve cusp
○ Left pulmonary arteries
○ Pulmonary artery
○ Pulmonary veins
○ Left atrium
○ Mitral valve cusp
○ Aortic semilunar valve cusp

○ Chordae tendineae
○ Papillary muscles
○ Left ventricle
○ Inferior vena cava
○ Right ventricle
○ Tricuspid valve cusp
○ Right atrium
○ Pulmonary veins
○ Pulmonary valve cusp

Figure 20-7

O RIGHT ATRIUM
O RIGHT VENTRICLE
O LEFT VENTRICLE
O LEFT ATRIUM
O Aorta
O Superior vena cava
O Right pulmonary veins
O Inferior vena cava

O Coronary sinus
O Right coronary artery
O Middle cardiac vein
O Great cardiac vein
O Circumflex artery
O Left pulmonary veins
O Pulmonary arteries

Figure 20-10

O Right coronary artery
O Cusps of the pulmonary
    semilunar valve
O Left coronary artery
O Coronary trigone
O Ascending aorta
O Cusps of aortic semilunar
    valve

O Tricuspid valve
O Anterior cusp
O Posterior cusp
O Septal cusp
O Interventricular septum
O Bicuspid valve

Figure 20-12

O Atria
O AV valve
O Chordae tendineae
O Papillary muscles
O Ventricles

O Cusps
O Aorta
O Pulmonary artery
O Semilunar valve

Figure 20-14

- ○ Superior vena cava
- ○ Right pulmonary artery
- ○ Aorta
- ○ Apex
- ○ Anterior interventricular artery, anterior descending artery
- ○ Left marginal artery
- ○ Circumflex artery
- ○ Left auricle
- ○ Left coronary artery

- ○ Left pulmonary vein
- ○ Left pulmonary artery
- ○ Pulmonary artery
- ○ Right pulmonary vein
- ○ Right coronary artery
- ○ Right auricle
- ○ Right anterior ventricular artery
- ○ Right marginal artery
- ○ Posterior interventricular artery, Posterior descending artery

Figure 20-16

- ○ SA node (pacemaker)
- ○ Right atrium
- ○ AV node
- ○ Right ventricle
- ○ Left atrium

- ○ AV bundle (of His)
- ○ Left ventricle
- ○ Right and left branches of the AV bundle
- ○ Purkinje fibers

162

Figure 21-7

○ Superior vena cava
○ Capillaries of the head neck and upper extremeties
○ Left pulmonary artery
○ Capillaries of the lungs
○ Left pulmonary veins
○ Left atrium
○ Left ventricle
○ Celiac artery
○ Capillaries of the lower extremeties
○ Arterioles capillaries

○ Venules
○ Capillaries of the pelvis
○ Capillaries of the gastro-intestinal tract
○ Hepatic portal vein
○ Capillaries of the liver
○ Capillaries of the stomach
○ Inferior vena cava
○ Right ventricle
○ Pulmonary trunk
○ Right atrium
○ Aorta

Figure 21-19

○ Right pulmonary artery
○ Right atrium
○ Aortic arch
○ Pumonary trunk
○ Left pulmonary artery
○ Left atrium

○ Left lung
○ Left pulmonary vein
○ Left ventricle
○ Right ventricle
○ Right pulmonary vein
○ Right lung

Figure 21-36

○ Pyloric vein
○ Hepatic vein
○ Inferior vena cava
○ Hepatic artery
○ Splenic vein
○ Pancreatic vein

○ Inferior mesenteric vein
○ Right colic vein
○ Superior mesenteric vein
○ Middle colic vein
○ Hepatic portal vein

Figure 21-42

- Wall of the yolk sac
- Blood islands
- Angioblasts (mesenchyme)
- Mesoderm

- Endoderm
- Lumen of the blood vessel
- Blood cells
- Endothelial lining

Figure 22-1

○ Right lymphatic duct
○ Right subclavian vein
○ Internal jugular vein
○ Cervical lymph nodes
○ Left subclavian vein
○ Thoracic duct
○ Mediastinal lymph nodes
○ Spleen

○ Cisterna chyli
○ Deep inguinal lymph nodes
○ Superficial lymphatics of the
    lower limb
○ Superficial lymphatics of the
    upper limb
○ Lymphatics of breasts
○ Axillary lymph nodes
○ Thymus

Figure 22-6

○ Afferent lymph vessels
○ Trabeculae
○ Hilus
○ Lymph nodules
○ Sinuses
○ Germinal centers

○ Medullary cords
○ Sinuses
○ Efferent lymph vessels
○ Vein
○ Artery

Figure 22-7

○ Retoauricular nodes
○ Occipital nodes
○ Superficial cervical nodes
○ Parotid nodes
○ Submandibular nodes
○ Deep cervical nodes

Figure 22-8 (a,b)

O Axillary nodes
O Supratrochlear node
O Lateral group of axillary nodes
O Apical group of axillary nodes

O Central group of axillary nodes
O Subscapular group of axillary nodes
O Pectoral group of axillary nodes

Figure 22-9

O Inguinal nodes
O Great saphenous vein
O Popliteal nodes

Figure 23-5

- ○ B cells
- ○ Antibodies
- ○ Competent B cells
- ○ Plasma cells
- ○ Viruses
- ○ T cells

- ○ Cytotoxic T cells
- ○ T cell receptor
- ○ Viral protein
- ○ Infected cell
- ○ Normal cell surface protein

Figure 23-8

- ○ Antigen
- ○ Antigenic determinants
- ○ Binding sites
- ○ Variable region
- ○ Constant region

- ○ Antibody
- ○ Heavy chain
- ○ Light chain
- ○ Disulfide bonds
- ○ Antigen-antibody complex

Figure 24-11

O Capillary endothelium
O Capillary basement membrane
O Interstitial space
O Alveolar epithelial wall
O Surfactant layer

O Alveolus
O Capillary
O Red blood cell
O Diffusion of oxygen
O Diffusion of carbon dioxide

Figure 24-15

O Inspiratory reserve volume
O Inspiratory capacity
O Vital capacity
O Total lung capacity

O Functional residual capacity
O Residual volume
O Expiratory reserve volume
O Tidal volume

Figure 24-23

O Thalamus
O Cerebellum
O Cerebrum
O Hypothalamus
O Pituitary gland
O Pneumotaxic area

O Midbrain
O Apneustic area
O Pons
O Inspiratory area
O Expiratory area
O Medulla oblongata

Figure 25-4

O Sphenoid sinus
O Frontal sinus
O Nasal conchae
O Hard palate
O Fauces
O Tongue
O Hyoid bone
O Vocal cord
O Nasopharynx
O Pharyngeal tonsil
O Posterior nares
O Opening of auditory
    (eustachian) tube

O Soft palate
O Uvula
O Palatine tonsil
O Oropharynx
O Lingual tonsil
O Epiglottis
O Laryngopharynx
O Larynx
O Trachea
O Esophagus

Figure 25-6

O Crown
O Neck
O Root
O Enamel
O Pulp cavity
O Artery
O Vein
O Nerve

O Spongy bone of alveolar
    process
O Root canal
O Cementum
O Ligament
O Dentin
O Pulp
O Gingiva

Figure 25-10

O Lesser omentum
O Esophagus
O Cardiac sphincter
O Cardiac orifice
O Greater omentum
O Fundus
O Body of the stomach
O Pyloric region of the
    stomach
O Greater curvature
O Rugae
O Tail of the pancreas
O Body of the pancreas

O Head of the pancreas
O Pancreatic duct
O Duodenal papilla
O Plicae circulares
O Hepatopancreatic ampulla
O Pyloric sphincter
O Duodenum
O Lesser curvature of the stomach
O Gallbladder
O Common bile duct
O Cystic duct
O Hepatic duct

Figure 25-21

- Mucosa
- Submucosa
- Muscularis
- Serosa
- Villi
- Epithelial cells lining the villus
- Brush border
- Lacteal
- Capillary network

- Goblet cells
- Nerve fiber
- Opening of intestinal glands
- Intestinal glands
- Vein
- Artery
- Lymph vessel
- Nerve

Figure 26-5

O Liver
O Site of Cellular respiration
O Site of Glycogenesis,
        Glycogenolysis
        and Gluconeogenesis
O Fat cells

Figure 26-7

O Fat cell
O Liver
O Other cells

Figure 27-3

O Juxtamedullary nephron
O Distal convoluted tubule
O Cortical nephron
O Capsule
O Proximal convoluted tubule
O Glomerulus
O Bowman's capsule

O Arcuate artery and vein
O Loop of Henle
O Collecting tubule
O Papilla
O Cortex
O Medulla

Figure 27-4 (a)

O Proximal convoluted tubule
O Parietal layer of Bowman's
   capsule
O Podocytes of visceral layer of
   Bowman's capsule
O Glomerular capillaries
O Fenestrations

O Afferent a.
O Juxtaglomerular apparatus
O Efferent a.
O Distal convoluted tubule

Figure 27-16

O Bowman's capsule
O Glomerulus
O Proximal convoluted tubule
O Distal convoluted tubule
O Vasa recta

O Loop of Henle
O Urea
O Collecting tubule
O Interstitial fluid

172

Figure 29-2

O Homologous pair
O Homologous pair forming tetrad

Figure 29-4

O Rectum
O Seminal vesicle
O Urethra
O Ejaculatory duct
O Prostate gland
O Anus
O Bulbourethral gland
O Pubis of pelvis
O Epididymis
O Scrotum

O Bladder
O Vas deferens
O Penis
O Cavernous bodies
O Urethra
O Testis
O Prepuce
O Glans penis

Figure 29-5 (b)

O Tail
O Spermatic cord
O Testicular veins
O Testicular artery

O Vas deferens
O Head of epididymis
O Seminiferous tubules

Figure 29-14

O Body of uterus
O Cervix of uterus
O Bladder
O Rectum
O Vagina
O Anus
O Vaginal opening

O Uterine tube
O Ovary
O Ovarian ligament
O Urethral opening
O Clitoris
O Labia minora
O Labia majora

Figure 29-15

O Opening of uterine tube
O Uterine tube
O Labia minora
O Ovarian ligament
O Endometrium
O Round ligament
O Ovary

O Body of uterus
O Muscular wall of uterus
O Cervix
O Vagina
O Hymen
O Labia majora

Figure 29-20

O Mons pubis
O Opening of urethra
O Perineum
O Vaginal opening
O Clitoris

O Labia majora
O Vestibule
O Hymen
O Labia minora
O Anus

Figure 30-3

○ Polar body
○ Blastomere
○ Zona pellucida
○ Degenerating zona pellucida

○ Inner cell mass
○ Blastocyst cavity
○ Trophoblast

Figure 30-4

○ Trophoblast
○ Inner cell mass
○ Inner cell mass
○ Trophoblast
○ Uterine epithelium
○ Uterine gland
○ Uterine blood vessel
○ Uterine epithelium
○ Chorion
○ Embryonic cavity
○ Healing site of implantation
○ Amniotic cavity
○ Maternal vessel
○ Amniotic cavity

○ Amnion
○ Embryo
○ Chorionic cavity
○ Chorionic villi (area of future placenta)
○ Embryonic stalk
○ Maternal venous pool
○ Chorion
○ Chorionic cavity
○ Amnion
○ Umbilical arteries and vein
○ Umbilical cord
○ Placenta
○ Amniotic cavity

Figure 30-5

○ Amniotic cavity
○ Blastocyst cavity
○ Exocoelomic membrane
○ Endometrial gland
○ Endometrial capillary
○ Syncytiotrophoblast
○ Embryonic epiblast
○ Cytotrophblast
○ Embryonic hypoblast
○ Blastocyst cavity
○ Cytotrophoblast
○ Syncytiotrophoblast

○ Amnion
○ Bilaminar embryonic disc
○ Amnion
○ Bilaminar embryonic disc
○ Extraembryonic mesoderm
○ Amniotic cavity
○ Maternal blood in lacunae
○ Cytotrophoblast
○ Primary yolk sac
○ Endometrial epithelium

175

Figure 30-7

O Maternal sinusoid
O Primary yolk sac
O Extraembryonic coelom
O Lacunar network
O Primary villus
O Extraembryonic somatic
   mesoderm
O Chorion
O Extraembryonic splanchnic
   mesoderm

O Maternal blood
O Remnant of yolk sac
O Primary villus
O Connecting stalk
O Secondary yolk sac
O Extraembryonic somatic
   mesoderm
O Surface epithelium

Figure 30-18

O Superior vena cavity
O Right atrium
O Ductus venosus
O Liver
O Portal vein
O Umblicial vein
O Umbilical arteries

O Aorta
O Ductus asteriosus
O Pulmonary artery
O Foramen ovale
O Left atrium
O Inferior vena cava
O Descending aorta